FOR IMMEDIATE RELEASE

CONTACT: Dominic Bria
Shingo Institute
Phone: 435-797-0771
Email: dominic.bria@usu.edu

"Building a Global Learning Organization" by Patrick Graupp, Gitte Jakobsen and John Vellema Receives Shingo Research and Professional Publication Award

Summary: After trained examiners and accomplished business professionals performed a thorough assessment of their book, Patrick Graupp, Gitte Jakobsen and John Vellema have been selected as recipients of an internationally recognized award from the Shingo Institute, part of the Jon M. Huntsman School of Business. The authors received the award at the 27th Shingo International Conference that occurred May 4-8, 2015.

LOGAN, Utah— The Shingo Institute, part of the **Jon M. Huntsman School of Business** at Utah State University, has awarded Patrick Graupp, Gitte Jakobsen and John Vellema with the Shingo Research and Professional Publication Award for their work "Building a Global Learning Organization: Using TWI to Succeed with Strategic Workforce Expansion in the LEGO® Group." Graupp, a two-time Shingo Research Award recipient and senior master trainer at the TWI Institute, collaborated with two practitioners who performed the planning and implementation of the LEGO Group's worldwide Learning Organization. Jakobsen is HR senior manager in the LEGO Learning Center and she leads LEGO's global training activities. Vellema served as project leader and concept developer of LEGO's Global Job Training Organization until 2012 when he started an advisory and training firm – *business through people.*

"Receipt of the Shingo Research and Professional Publication Award signifies an author's significant contribution to advancing the body of knowledge regarding enterprise excellence," said Mark Baker, executive director of the Shingo Institute.

"Building a Global Learning Organization" describes how a multinational company developed a global structure for learning based on the TWI (Training Within Industry) program to create and sustain standardized work across multiple language and cultural platforms. The book outlines the organizational and planning models used by the LEGO Group to create the internal ability to give and receive tacit skills and knowledge. Describing how and why TWI is used as the foundation for success in knowledge transfer across diverse languages and cultures, it provides step-by-step guidance on how to establish a solid organizational foundation for your own Learning Organization.

Jeffrey K. Liker, professor at the University of Michigan and Shingo Research Award-winning author of "The Toyota Way" says, "LEGO has been a household name all of my life, and I was aware they had a strong people-focused culture and adopted lean methods. This book, written with LEGO insiders, is a stunning example of the discipline and commitment needed to develop people as masters of their crafts through the only way people learn—repetitive, deliberate practice." And John Bicheno, founder of MSc in Lean Enterprise at The University of Buckingham says, "Few books, if any in the lean area since the NUMMI era, have gone into such depth on what it takes to integrate and unify across cultures. The book will become a standard guide not only to TWI implementation, but to the wider challenge of cross-functional and cross-cultural integration."

By "challenging" or applying for an award, authors invite a group of accomplished professionals and trained examiners from the Shingo Institute to thoroughly review their publications. The examiners select recipients based on a rigorous set of standards.

The authors received the award during the Awards Gala of the 27th Shingo International Conference held the week of May 4-8, 2015. The conference is a five-day event featuring a selection of workshops, plant tours, keynote speakers and breakout sessions designed to provide ongoing knowledge, insights and experience for organizations in their pursuit of operational excellence.

About the Shingo Institute

Housed at the Jon M. Huntsman School of Business at Utah State University, the Shingo Institute is named after Japanese industrial engineer Shigeo Shingo. Shingo distinguished himself as one of the world's thought leaders in concepts, management systems and improvement techniques that have become known as the Toyota Business System. Drawing from Shingo's teachings and years of experience working with organizations throughout the world, the Shingo Institute has developed the Shingo Model™ which is the basis for several educational offerings including workshops, study tours and conferences. It also awards and recognizes organizations that demonstrate an exceptional culture that continually strives for improvement and progress. Those interested in more information may visit **www.shingo.org**.

Patrick Graupp / Gitte Jakobsen / John Vellema

Building a Global Learning Organization

Using TWI to Succeed with Strategic Workforce Expansion in the LEGO® Group

CRC Press is an imprint of the
Taylor & Francis Group, an **informa** business
A PRODUCTIVITY PRESS BOOK

CRC Press
Taylor & Francis Group
6000 Broken Sound Parkway NW, Suite 300
Boca Raton, FL 33487-2742

Printed on acid-free paper
Version Date: 20140320

International Standard Book Number-13: 978-1-4822-1363-8 (Paperback)

Library of Congress Cataloging-in-Publication Data

Graupp, Patrick.
Building a global learning organization : using TWI to succeed with strategic workforce expansion in the LEGO Group / Patrick Graupp, Gitte Jakobsen, John Vellema.
pages cm
Includes bibliographical references and index.
ISBN 978-1-4822-1363-8 (pbk. : alk. paper)
1. LEGO koncernen (Denmark)--Management--Case studies. 2. Employees--Training of--Case studies. 3. Organizational learning--Case studies. 4. Organizational effectiveness--Case studies. I. Title.

HD9993.T694L443 2014
658.3'124--dc23 2014007817

Visit the Taylor & Francis Web site at
http://www.taylorandfrancis.com

and the CRC Press Web site at
http://www.crcpress.com

Love. Faith. Devotion … and Patience. These are gifts that when freely given will be returned in full measure. For all that they give to us, we dedicate this book to our families and loved ones: Arden, Reilly, Tyler … Peter, Julie, Sarah … Maria.

We also dedicate this book to all our great LEGO colleagues who have been part of the Learning Center and our TWI journey. Without you we would never have reached this achievement.

Contents

SECTION III CREATING THE GLOBAL LEGO TRAINING ORGANIZATION

Foreword

This book is a case example of one of the most fundamental qualities of human nature, namely, the passing on of knowledge.

The giving and receiving of knowledge is a natural and vital part of being human. Depriving people of this basic need has a detrimental effect on a young person's psychological development. Everybody who has had the luck of having a good teacher, coach, or parent knows the importance of this and is grateful for the fact that they were blessed with having this person to help them learn. Especially in our working lives, the ability to learn and develop is vital for each individual as it is the only insurance a person can have of attaining a successful career. Also from a company perspective, knowledge management is considered to be the most important source of a company's sustainable competitive advantage. Yet we are still far from understanding the method by which we pass on knowledge in an optimal way.

Nearly all large companies are setting up manufacturing processes in new locations either through geographic expansion or footprint reorganization. Companies have done this with varying and variable degrees of success. Even within the same company, some attempts may be very successful while others are less so. It is usually not possible to forecast which will be successful and which will be problematic but one thing is sure, that no company will be successful if it doesn't understand the importance of knowledge management and invest in efforts to ensure that transference of knowledge is possible, controlled, and supervised.

The challenge, of course, is the transference of tacit skills and tacit knowledge. This book is one story of the experience and approach that we took at the LEGO Group to ensure the transference of tacit knowledge and tacit skills in order to expand our capabilities. Tacit knowledge and tacit skills have been passed down from one generation to another from master to apprentice for centuries. Connoisseurs, craftsmen, and tradesmen all know this art of learning. But times have changed and the conditions for passing on this kind of knowledge have also changed, yet the human brain and body have not changed.

The common distinction between tacit *skills* and tacit *knowledge* is that tacit skills come from practice and tacit knowledge comes through the combination of explicit knowledge domains in a dynamic thinking process. Tacit knowledge is typically seen in design processes where engineers or the like apply tacit knowledge to come up with innovative designs. Tacit skills are typically seen when these designs are manufactured through the application of craftsmanship. It is when tacit design knowledge and manufacturing skills are combined in synergy that new and innovative knowledge is created.

A learning organization is an organization where employees have the ability to synthesize tacit and explicit knowledge in their working environment. It is only by recognizing the appropriateness and contextual value of tacit knowledge that the application of knowledge management adds value. In a production environment, it is only by way of this knowledge that organizations start to learn the concept and implications for actually managing knowledge in a systematic and deliberate way.

This book is about how we applied the principles of working with tacit knowledge and skills when transferring knowledge to and between our manufacturing sites. When we introduced the trainer program at the LEGO Group, we were very conscious of ensuring that we be loyal to the basic principles when working with tacit knowledge and tacit skills. From respect for the knowledge to be attained to the effort and practice needed to be a master, we designed the training accordingly. Any trainer who was chosen was made aware of the importance of understanding the guiding principles that were required to be a good trainer.

Training Within Industry (TWI) was a big part of the success of this approach and that is how this book came about. It actually started with some seminars that became very popular and Gitte, John, and Patrick were encouraged to write this down. Thus the birth of this book.

This is a very small part of a bigger journey that we are on at the LEGO Group to create a learning organization. We are still at the infant stages but I am convinced that by writing this book we can further develop our method and for me that is the main motivation. I am proud when I read some of the quotes from some of the trainers on how the training has influenced them as human beings. I feel we are making a small contribution to the development of a new generation of the master/apprenticeship method that has been wanting since the disappearance of craftsmanship and the industrialization of work.

Bon Voyage
Stephen J. Burke
Vice President Operations HR
The LEGO Group

Acknowledgments

We were incredibly fortunate to assemble and work with a group of talented and passionate people in the LEGO organization and would like to express our appreciation to the core project team in the pilot project for the commitment they have shown to the Learning Center project and their contributions to the completion of this book.

From Mexico: Gustavo Otero, Alberto Rosas, Marien Perez Vertti, Aydeli Rios, and Ericka Hernandez
From Hungary: Klaus Alexander Nousiainen, Fruzsina Veress, László Zsíros, and István Nagy
From the Czech Republic: Pavel Kroupa
From Denmark: Marcus Haure, Flemming Tiro Lund, Bent Jensen, Tommy Elkjær Højsager, Kent Kjærhus, Jacob Møller Lund, Johannes H. Lystbæk and Ole Therkelsen

In addition to the core team, there is no better assembly of enlightened leadership than can be found in the LEGO Group who worked with this project to set direction, give support, and ensure a successful completion. We would like to express our sincerest appreciation and respect for the following individuals.

Stephen Joseph Burke, Jens Peter Clausen, Jes Bladt, Anders N. Ravnskjær, Laszlo Madura, Henrik P. Sørensen, John Hansen, Csaba Toth, Beatriz Perez, Charlotte Bekker, Zuzana Chapmanová, and Merete Laursen

We would like to thank Bob Wrona and all the dedicated staff at the TWI Institute who provided invaluable guidance and support to make sure the TWI methodologies were properly introduced and implemented.

In particular, Lynne Harding worked tirelessly supporting the translations of the training manuals and materials into the various languages, no easy feat.

We would also like to thank Peter Enevoldsen, for his assistance in preparing parts of the draft; also, Laura Mott and John Frost for their coaching and support.

Finally, we would like to thank our editor Michael Sinocchi who believed in and supported this project wholeheartedly from day one and our project editor, Iris Fahrer, for whom no detail was too small to pursue. Without you two, we would not be here.

About the Authors

Patrick Graupp began his training career at the SANYO Electric Corporate Training Center in Japan after graduating with highest honors from Drexel University in 1980. There he learned to deliver Training Within Industry (TWI) and other training programs for SANYO employees inside and outside of Japan. He was transferred to a compact disc manufacturing facility in Indiana where he gained manufacturing experience before returning to Japan to lead SANYO's global training effort. During this time, Graupp earned an MBA from Boston University and was later promoted to head up human resources for SANYO North America Corp. in San Diego, California.

Graupp partnered with Bob Wrona in 2001 to conduct TWI pilot projects in Syracuse, New York that became the foundation for the TWI Institute which has since trained a vast network of certified trainers who are now delivering TWI training in the manufacturing, health care, construction, energy, and service industries around the globe. These efforts are outlined in their book *The TWI Workbook: Essential Skills for Supervisors* (CRC Press, 2006), a Shingo Research and Professional Publication Prize Recipient for 2007. Graupp also authored *Implementing TWI: Creating and Managing a Skills-Based Culture* published by Productivity Press in 2010, and *Getting to Standard Work in Health Care: Using TWI to Create a Foundation for Quality Care* published by CRC Press in 2012.

Gitte Jakobsen has been involved in organizational development and knowledge management in the LEGO Group since 1997 with roles as both staff manager and project manager in the LEGO marketing arena and later within LEGO Operations. She has extensive experience in the development of complex marketing and production processes based on her experience

setting up a LEGO marketing development and production function in the Czech Republic and her activities creating Learning Centers in LEGO production sites in Denmark, Mexico, Hungary and the Czech Republic.

In 2009, Jakobsen completed a Master's degree in educational psychology at the University of Aarhus and has since been working as a Learning Specialist in her position as HR senior manager in the LEGO Learning Center, providing both deep practical and theoretical perspectives. Her core responsibilities lie in leading global activities while building up training organizations and knowledge management activities across LEGO production and engineering functions including a new LEGO factory in China. Additionally, she is developing global LEGO programs like *World Class Craftsmanship*, with the objective of building up LEGO toolmakers, and *Technology Leaders of Tomorrow*, both initiatives focused on developing standardized capabilities across cultures and LEGO sites.

John Vellema started his career as a toolmaker in injection molding. This deeply rooted experience working with and understanding life on the shop floor provided him with invaluable experiences that cannot be learned in a classroom. He also served in the Danish Army where his last position was First Class Sergeant for a recon unit stationed in Kosovo. After leaving the Army, Vellema earned an engineering degree in manufacturing and management from the University of Southern Denmark. He then joined the LEGO Group in 2007 participating in the Supply Chain Graduate Program. In combination with years of coaching teams under the Danish Association of Rowing, this positioned him to become project leader and concept developer over the next three years, beginning in 2009, in the development and implementation of the Global Job Training Organization at LEGO. Through these experiences he has gained a strong understanding of the link between production, leadership, and engineering.

In 2012, John left the LEGO Group to start an advisory and training firm, *business through people*. The firm's primary objective is to help companies create and maintain a highly skilled and motivated workforce. *business through people* has since become a recognized company supporting businesses and corporations across Europe.

Introduction: The LEGO Group and Its Roots

It goes without saying that the LEGO Group is one of the world's most iconic brand names whose products have been loved by children, and adults, from one generation to the next since it first patented the "LEGO brick" in 1958. By the end of the millennium, the LEGO brick was named Toy of the Century by the British Association of Toy Retailers and one of the Products of the Century by *Fortune* magazine. From humble beginnings in the small town of Billund, Denmark, Ole Kirk Kristiansen, and then his sons and grandchildren after him, transformed a small carpentry shop into what would be, in 2013, the world's second-largest toy company. Over the decades, the company grew steadily, although not without some bumps along the way, and by the 2010s, it was poised for yet another period of growth, which would see the company rapidly expanding its sales and manufacturing overseas.

Ultimately, this rapid expansion would lead it to look carefully at its global workforce, which would be, in the end, the key to its ultimate success. With a long history of valuing the educational development of children who play with its products, LEGO management would now need to look inward at their own workforce and ask how they could, in both word and in action, become a true global learning organization. Education and training would be critical elements of this transformation. More specifically, as the expansion gained steam and restructuring placed more emphasis on global operations, the growth of the workforce outside Denmark would create the need for better skills in handling a multicultural employee base.

This is where we, the authors, come in. Gitte Jakobsen and John Vellema were both key members of the LEGO Operations team that gave birth to a new training organization that would "break the code" of how to effectively create and sustain standard work processes across international borders, languages, and cultures. They led the two parts of the pilot team within operations and human resources, and together with their Global Learning Center colleagues, they would transform how LEGO employees around the world learn to do their jobs in a correct and consistent way *after* working together to find the *best way* these jobs should be performed by everyone. Patrick Graupp was the leading expert on the Training Within Industry, or TWI, program that they adopted as the core methodology for this new learning structure. He assisted and guided the team in mastering the use and deployment of this tool within LEGO Group operations. Together, we tell the story of how the Global LEGO Training Organization was conceived, developed, and deployed. Since all three of us were participants in the process, from this point we refer to the actors in this story as "we."

In searching for tools that could create standard work on an international scale, we found the TWI program after an extensive evaluation of what was needed for the company to create global standards so that work could be done anywhere in the world with consistent quality that met customers' demands and expectations. TWI provided the basic skills we felt were needed to enter the front lines of the company's operations and directly touch and affect the people who put their hands on the product every day to ensure a perfect delivery. But, before simply introducing a new tool to the organization, the bigger piece would be to create an organizational structure that would support the use of this tool correctly and continuously.

This book is about how the LEGO Group created a global learning organization in operations as a vital part of the company's effort to ensure the future viability of the business. It is a story of patience, perseverance, and cultural empathy in creating culture change that would span the globe while keeping true to the core values of one of the world's most beloved companies. Most of all, though, it is an account of how the needs of organizational renewal in the twenty-first century could be made using a tool that contained age-old wisdom and understanding from the World War II era.

First, let us look at how the company got to where it is today.

The Incredible LEGO Group

No discussion of the LEGO Group and its signature product, the LEGO brick, can begin without first marveling at the incredible impact this toy has had on our culture. The company, from its earliest days after patenting the LEGO brick and creating a System of Play in which a variety of parts could be assembled in countless ways and then disassembled again, focused intently on the development of creativity in children through the fun use of their product. Indeed, the stated aim given for the company's products is to create learning and development opportunities for children to "inspire and develop the builders of tomorrow" through play.*

The ability of LEGO bricks to achieve this lofty goal can be readily seen when considering that the company produces approximately 4,200 different elements in 58 different colors, and all of these elements conform to a singular system in which any element can be connected to another one in an infinite number of combinations. Even if we look at just one of these elements, the basic eight-stud (2 × 4 rectangular) brick, mathematicians have calculated that six of these bricks of the same color can be assembled in over 915 million possible combinations—915,103,765 to be exact. The challenge of building complex assemblies or creating new structures from scratch is an ideal way, according to scholars at the Massachusetts Institute of Technology (MIT), for children to learn to think systematically and creatively.†

In their 2012 Progress Report (available online at the LEGO website), the company defined its core product:

> The LEGO brick has qualities that allow you to be artistic, scientific, intuitive and deductive, tactile and analytical all at the same time. In a safe environment, the bricks provide play experiences that make you comfortable with uncertainty and able to frame a problem. These are the skills of those who will build a better tomorrow.

Even while staying focused on ideals around creative learning and development for children, or perhaps by staying focused on these ideals, the LEGO Group has built a strong record of sales and customer loyalty for the core product line with which few companies can compete. Most significantly, it has been achieved with strong profitability. In 2012, their net profit margin (net profit/revenue) was 24.0%.‡ Total revenue has expanded four times since 2004, and importantly for our story, the average number of

---

* Company Profile 2012, p. 4, cache.lego.com/r/.../ts.20120420T123711.Company_profile_uk.pdf.

† Delingpole, James, When LEGO lost its head—and how this toy story got its happy ending, *Daily Mail*, December 18, 2009, http://www.dailymail.co.uk/home/moslive/article-1234465/When-Lego-lost-head--toy-story-got-happy-ending.html.

‡ LEGO Annual Report 2012, LEGO Annual Report 2008, http://www.lego.com.

full-time employees, while holding steady at around 5,000 people from 2004 to the end of 2008, doubled to 10,400 by 2012.* This number is projected to continue growing in the coming years as the company expands its production capacity into Asia.

Some interesting facts help us to get a feel for the breadth and scale of what the LEGO Group has achieved:

- There were 45.7 billion LEGO bricks (elements) produced in 2012; that was 87,000 bricks every minute or 1,450 every second.
- There are 80 LEGO bricks for every person in the world.
- To date, approximately 600 billion LEGO bricks have been produced.
- LEGO products are sold in more than 130 countries.
- Laid out end to end, the number of LEGO bricks sold in 2011 would reach more than 16 times around the globe.
- Eight LEGO sets are sold each second, and during the Christmas season, that number increases to 28 sets sold each second.
- With a production of over 300 million tires in 2011, the LEGO Group is one of the world's largest "tire companies."
- Over four billion minifigures have been produced, making this the world's largest population group.
- A LEGO brick manufactured in 1958 still fits snugly into a brick made in 2013.

History of the LEGO Group

In 1916, Ole Kirk Kristiansen, a Danish carpenter, bought a workshop in the small town of Billund, which lies in the center of Jutland, the landmass that protrudes north of Germany but is culturally and historically part of the Scandinavian countries directly to the north and east across the North Sea. Ole Kirk Kristiansen ran his business building houses and furniture, but when the Great Depression threatened to put him out of business, he turned, in 1932, to making wooden toys such as yo-yos, which were popular at the time, and pull-along animals and vehicles. His motto was "Only the best is good enough," and he named the company LEGO, which is the composite of two Danish words, *LEg GOdt* meaning "play well." Many years later, it was learned that the word in Latin means, "I put together."

---

* Ibid.

After the Second World War, in 1946, the company bought a plastic injection molding machine, the first of its kind in Denmark, and by 1949 was producing some 200 different plastic and wooden toys, all for the local Danish market. These included what they called Automatic Binding Bricks that they designed by modifying a sample plastic toy building block they had received with the molding machine from England. These would be the precursors of the LEGO brick we know today. In 1953, they would be renamed *LEGO Mursten* or "LEGO bricks" and would soon thereafter be marketed as a revolutionary System of Play, with toys no longer "one-offs" but integrated around a theme, like a town plan with buildings, streets, cars and trucks, and so forth. These could then be sold separately as individual sets or accessory kits.

In 1957, they invented a new way of interlocking the bricks by molding tubes inside the bricks that would better hold, when two bricks were snapped together, the studs of one brick between the tubes and the inner sides of the second brick. The original brick style had the studs on top, but the inside of the brick was a hollow rectangular cavity. So, the new stud-and-tube coupling system made the models built with LEGO bricks much more stable. The company patented this new coupling system on January 28, 1958, and this became the start of the modern LEGO era.

Beginning in the 1960s, production of wooden toys ended, and they concentrated all their efforts on promoting the LEGO System of Play. Throughout the decade, the variety and size of the LEGO sets grew considerably. By the end of the decade, the first LEGOLAND theme park was opened in Billund, and the LEGO DUPLO series for very young children was introduced, with the bricks twice the size of the standard LEGO brick for easier use and safety. Going into the 1970s, the company introduced the minifigure, an innovative addition so that characters and story lines now became an integral part of the child's experience. The LEGO Technic series was also launched as a way of getting older children interested by offering them more challenging constructions of vehicles and other machines.

The 1980s and 1990s continued to see steady growth and a dizzying array of new themes. In 1998, the LEGO Group entered into licensing agreements with Lucasfilm Limited to develop, produce, and market products based on the *Star Wars* theme, one of the most successful product lines in the company's history. Similar agreements were made with Disney, Warner Bros., and other movie and entertainment companies. As noted, the company's products ended the millennium being named the Toy of the Century. Moving into the new century, the company introduced a successful new series of products for girls, LEGO Friends, which was a feat they had been trying to achieve for decades since the vast majority of its customers have always been boys, and the Ninjago series based on a ninja theme that has become a worldwide sensation. Along with substantial organizational changes to make the LEGO Group more adaptable and competitive, including the learning organization we describe in this book, the company is poised to continue its journey toward sustained future growth.

How to Read This Book

Typically, a story is told in a linear fashion: You begin at the beginning and keep going forward, systematically, until you come to the end. However, in telling the story of how the LEGO Group developed the Global LEGO Training Organization, there were many moving parts that, while occurring simultaneously, cannot be explained clearly or concisely at the same time. For example, even as the program details were in development early in their conceptual stages, efforts were already under way to secure across-the-board acceptance and support from key stakeholders throughout the worldwide organization. And, even as a pilot project was initiated to demonstrate the effectiveness of the new methodology, the project team was still creating and adjusting the organization and its tools to find the right balance and fit that would continue to work long after the pilot was completed.

So, we have divided the book into sections to explain the different aspects of the entire project: laying the groundwork, the pilot project, organization/tools/methodology development, and rollout. Here is a brief overview of the sections to help you read them effectively.

Section I: Laying the Groundwork

Section I describes the "burning platform" that created the need for a new training structure and the requirements it would have to fill. It reviews the initial steps taken to show how and why we arrived at our final organizational structure and what new roles and functions we would need to create to reach that end. Having conducted an extensive needs analysis, we review those results and the key problems and concerns that came up with the state of training at that time. This led to the design of a pilot project to prove our new theories and methodologies; we outline here the key components and design of that project. We discuss the major tools we would deploy in the pilot, in particular the TWI Job Instruction program and why it was selected to be the foundation of the program. We present the argument for why creating standardized work throughout the manufacturing processes in all LEGO plants around the world would be the goal of the new training organization. Finally, we give a high-level overview of the full development model that we would pursue to achieve these lofty goals.

Section II: The Global Pilot Project

In Section II of the book, we move away from the theory and delve into the practicalities of what it took to actually make the pilot project run successfully. We discuss in detail how the people involved truly felt about what they were being asked to do, the challenges they encountered, as well as what they did to overcome them. We go into the details of how we ran the pilot as a series of workshops and look at the development and growth of the pilot project team over a 6-month period. By seeing what changes and adjustments were made along the way, we show how any culture change as large as the one we were trying to make cannot be simply a plan on paper but must be a process of full engagement with careful attention to the people involved, their feelings, and their fears and expectations. We describe also how the tools and processes we were trying to demonstrate in the pilot were used and what we learned about them along the way to final approval of the project for wider rollout throughout the company.

Section III: Creating the Global LEGO Training Organization

To this point in the book, we have introduced the various roles and tools that make up the new training organization we created and how we proved them in the pilot project. In Section III, we go into details that are more analytical, showing models and giving structure that is more concrete to the individual aspects of how the program works. For example, we give the content of the personality profiles we look for in each of the new trainer roles and explain fully the process of how we select and develop people to fill these roles. We describe the training programs we give to these new trainers so they will be prepared to execute their new assignments successfully. Then we delineate the plan, along with the tools and methodologies, to be used in moving the new organization forward so that it can execute effective training on an ongoing basis.

Section IV: Rollout

In the final part on the rollout in Section IV, we return to the more practical experiences of how we moved beyond the pilot project to program management that would sustain the effort and make the new training organization a regular part of how the work was achieved in the LEGO Group. In the end, this was the most difficult challenge, and we give examples of key decisions and actions that made this a success. Then, we introduce some of the opportunities that came later, leveraging the success of the training structure now in place. You will also discover how the new learning organization is used to help prepare for a brand-new manufacturing facility in China. We finish with some feedback from a few of the key participants in the process and how they saw this experience.

Keep in mind that, while reading any one section, there will be critical content and details that will be only lightly touched on or even skipped over because that discussion and explanation are done more thoroughly in another section. You may choose to read the sections out of order. If, for example, you are mostly interested in the content and theoretical underpinnings of the programs we implemented, read Section III first and come back later to see how we developed and implemented those methods. On the other hand, if you most desire to learn how we worked with the people to effect a major cultural change across multiple countries and languages, you should start with Section II. Begin with Section I if you want to see first why we needed to create a new training organization and how we came up with the conclusions that we did.

I

LAYING THE GROUNDWORK

Chapter 1

Setting the Course

Introduction

The LEGO Group started out as a small carpentry shop with a proud tradition of craftsmanship in Billund, Denmark, where everyone, under the direction of the founder and owner Ole Kirk Kristiansen, did their utmost to keep up with production and sales. As we saw in the book's Introduction, this business grew by leaps and bounds after the adoption of plastic injection molding in the early 1950s and the development of the LEGO System of Play based on the LEGO standard brick, plastic elements that could be fitted together in an infinite number of designs and shapes. While the business developed and expanded over the years, though, it was difficult to transition from a craftsmanship-based production system to a more global mass production system with standardized work procedures. You cannot simply transfer across borders and cultures all the learning and experience of many decades of in-house development of unique LEGO operational know-how. Understanding the complexity of finely tuned crafted machinery and the way one worker knows exactly how to approach an assignment and make it work is not something easily learned in a short time.

In 2005–2006, the company embarked on another growth spurt with a vigorous expansion focusing on the company's core products (e.g., LEGO City and LEGO Star Wars). This growth, however, would require more production outside the original Billund workshops at international plants worldwide; a focus on standardizing processes would be necessary to control the production. Efforts at reducing the complexity and number of LEGO elements and focusing on supply chain planning were a strong start at managing this situation. LEGO planners carefully went through the entire list of elements and product units, paring them down to fundamental structures that any child could understand and removing all the frills and extras that had enamored the company's adult designers for many years while still giving children what they really wanted from our products. Nevertheless, a more fundamental change in the culture of the company would be needed to make the kind of leap forward the LEGO Group wanted.

The Need for Change

With the company poised for solid growth, human resources (HR) staff within LEGO Operations looked at existing methods of *knowledge transfer* and *training over time*; then, we conducted a comprehensive needs analysis to shed light on the business issues that needed to be addressed by the company and our workforce. The contents of the needs analysis (more a series of discussions with LEGO employees at all levels of the organization in plants around the world than a formal survey or questionnaire) is discussed in detail in the next chapter. But, the fact that we spent the better part of half a year obtaining honest and comprehensive feedback from the production sites set the stage for what was to become a major overhaul in how the company provided training to its frontline employees.

Based on our conclusions from the needs analysis, the resulting proposals outlined the need for building and expanding workforce capabilities. Moreover, with the company's renewed growth targets, the worldwide workforce was due to expand tremendously over the following several years. This led to the search for and discovery of tools that would provide the needed changes; here, a high focus was placed on creating a LEGO way of building a global job training organization that would be capable of training the workforce of the future. We realized that to maintain long-term future viability we would have to dig deep into our culture and revitalize the method and structure by which we performed job training. We also came to the realization that this would entail re-creating ourselves to become a global learning organization as part of building a sustainable platform for growth.

Early in the new decade, the company announced its goal for expansion on a global scale, which included bringing on board as many as 7,000 to 8,000 new employees in the following 2–3 years. The company had already almost doubled the number of full-time employees from 2007 to 2010 (see Figure 1.1). Understanding the magnitude and full impact of

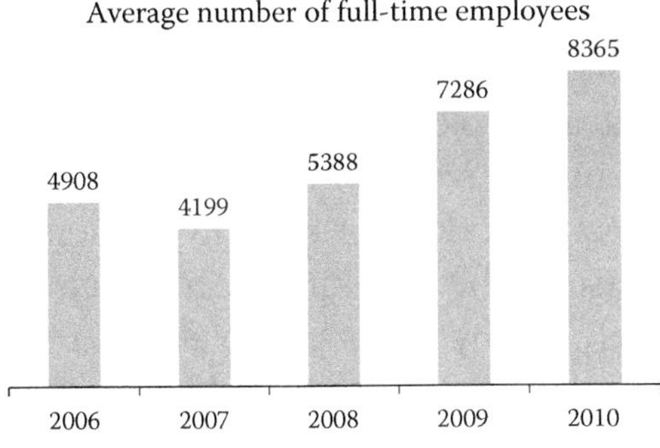

Figure 1.1 Average number of full-time employees.

the proposed expansion was to be the first essential step to our realizing the necessity for developing a structured job training organization. To be able to share and develop LEGO operational knowledge and skills in a systematic way with the expanding members of the worldwide LEGO family, some organized methodology of having a diverse collection of peoples of different languages and cultures working together in a stable, standardized fashion would be needed. It had been known for years that this worldwide expansion was part of the long-term strategy of the company. But, when we started the project, we did not know the actual numbers we would need to prepare for the growth.

Strategic Capability Building

To meet the demands of the growth strategy, a huge global expansion of capabilities was needed in the LEGO Group. This meant that the production sites were going to be scaling up; at the same time, these production sites would be required to perform processes with a higher level complexity than they had experienced until then. To handle both the growth and the complexity, it would be crucial to develop and improve local capabilities to make the production sites self-sustainable and capable of performing both plastic injection molding operations, which form the LEGO bricks, and mold construction processes, which create the molds from steel blocks, typically called tool and die making in the American context.

Based on these needs, it was important to work strategically and proactively with each site to ensure that we had the right people at the right location at the right time and that cost was in alignment with business demands. The way we did this was by defining strategic capabilities and capacity requirements by location based on business priorities for both the long and short terms. Because of the nature of our product (plastic elements of varying shapes, sizes, and colors), molding capabilities are one of the strategic areas of the company's capabilities that must be well established and standardized in all our production sites. Even though the LEGO brick may look simple, because of the high-quality requirements of the bricks, which must attach snugly to any other LEGO element built over the last six decades, they have to be produced using skillful techniques and processes. In addition, because of the high manufacturing volume, capacity requirements must also be defined to determine the number of people needed in the different positions at each production site.

By creating an overview of the strategic capabilities and the capacity requirements, we obtained a full picture of resource requirements by considering strategic as well as core capabilities. Here, we made specific plans of what it would require to ensure that the necessary capabilities involving **work** (strategy, objectives, priorities); **people and stakeholders** (skills, experience, motivation); **formal organization** (structure, tools, processes); and **culture** (history, beliefs, values) were developed and became available at the production sites. A planning tool used for creating this optimal mix was called *Build-Buy-Borrow*. This created an overview of how many people within each position we should *Build* (train and develop), how many people we should *Buy* (recruit), and how many people we should *Borrow* (e.g., expatriates, or trainers we brought in from an existing or more experienced site to support the development process). This provided a clear picture of what was required either from a new production site or from a more experienced production site that needed help in the process.

To support this approach, a formalized strategic workforce planning process must be part of each business area's planning function. This was primarily a leadership planning tool to ensure alignment between the LEGO Group strategy and related business plans, priorities, and organizational capabilities as explained in the previous paragraph. The LEGO Group defines a *core capability* as the key enabler of strategic activities and the foundation for our competitive advantage. In other words, capabilities contain more than just competencies. They are both the "soft" elements of an organization in the form of individual skills, competencies, and culture, and the "hard" elements in the form of organizational processes, structure, and systems. This was one of the areas that had been a challenge to instill in the company's leaders because they only considered skills and competencies, but not culture and processes. They looked only at the individuals within the organization when they considered competencies and not the organization as a whole, which is a system of processes reflected by the culture—the sum total of the human behaviors of those individuals. What they needed was a wider vision of "capabilities."

To construct a framework for systematic capability building, we looked at the four different areas mentioned: work, people, culture, and formal organization. We called that our congruence model, supporting the holistic and the systematic execution of the maintenance and development of core organizational capabilities. Furthermore, the ultimate purpose of the model was to identify key stakeholders for each core capability. By working with the capability building model and our strategic workforce planning, we moved from knowing little about the real essence of our workforce to clearly understanding the various global characteristics and situations as well as important local differences.

Building the Molders of Tomorrow

What we saw as core capabilities were those capabilities that the LEGO Group performs uniquely well, and that are of strategic importance to maintaining our status as a world-class toy manufacturer. We looked at these capabilities as *above* what is best in the industry or critical to simply being a viable player. These, in turn, are built on foundational capabilities that any company must have simply to participate in the business.

Our core capabilities have been defined as, for instance, product innovation and, in line with that, our capabilities in mold production and molding. To maintain and develop these core capabilities while supporting our aspiration to become a capabilities-driven company, it was important that we continuously improve and build on these core capabilities to ensure they remain unique to the company. The importance and the priority of core capability were best described in our global engineering and quality and manufacturing strategy:

Build and Maintain Core Capabilities

> "Mold and molding" are core capabilities. It is crucial to capture, develop, and protect our knowledge in molds and molding (the secret of how to make LEGO bricks). The clear target is to become the best in the world as measured by innovation and cost. Thus, we always want to maintain and develop our unique knowledge in the manufacturing system. We will always prioritize hiring the best people and overinvesting in the development of their capabilities to maintain the competence advantage of the company.

The purpose was, therefore, to create a solid framework with a coherent approach that identifies, maps, and develops capabilities systematically. When starting up the pilot project in the molding area, the incentive was to build capabilities in a deliberate and systematic way to support the journey toward strategic capability building, just as all new projects and initiatives are based on supporting the LEGO strategy, business plans, and other strategic initiatives.

"And Notice Nothing Else but the Language and the Local Temperature"

A clear direction to initiate activities supporting the development of the workforce was set right from the start by the LEGO top management:

> Any employee of the LEGO supply chain should be able to move from any factory to any other factory and notice nothing else but the language and the local temperature.
>
> And any employee of the LEGO supply chain should be open and respectful towards other cultures and learn how to best materialize our vision of Strength Through Diversity.

This vision statement expressed the core of what became a standard view of global expansion, and it gave a strong direction that was uniquely powerful when communicated on all levels. It would also prove to be the "guiding star" for creating the Global LEGO Training Organization for professionals and practitioners in LEGO Operations.

Growth Creates Need for Training

Although there were, of course, many factors involved in the kickoff of the Global LEGO Training Organization in operations, the main reason for developing this new part of the worldwide organization was the need for large growth in global operations of the LEGO Group that would support the anticipated growth in sales. The global footprint was growing fast with the steady increase in sales, and it was well understood that this expansion would not slow any time soon. So, LEGO Operations had to expand quickly to keep up. By creating a learning system that could improve the sharing of LEGO Operations' knowledge and skills between coworkers locally and globally, we would comply with the overall growth strategy.

As pointed out previously, there had always been a proud and strong tradition of craftsmanship in LEGO Operations going back to the early beginnings of the company. In this tradition, some of the good routines and procedures were documented, but never in a systematic and "trainable" way. Rather, they were passed on through word of mouth or in written documents, like handouts, with no formal training. This was something that needed to be changed for a global rollout to succeed.

We were first made aware of this gap in LEGO operational technique when the company tried to outsource some of this tacit know-how to another company. It was revealing when we first recognized for ourselves the true complexity of the production of our LEGO bricks and how much we underestimated the difficulty of transferring what was still essentially a craftsman-based process lacking standardization. How do you actually share all of this tacit know-how, the "crown jewels" of the company that have been gathered by a few craftsmen over many decades? The language and the temperature were at this time not the only things noticed by the employees who went overseas to pass on their knowledge and skills.

This is when Stephen J. Burke, the head of HR in Operations, was given the task of developing a global training system to handle sharing of production knowledge and skills.

Laying the Foundation for a Global Learning Center

Having neither a department nor a single employee to handle this demanding task, Burke started by looking at internal employees who had skills in organizational development and change management as well as in Lean and project management. He also wanted people who had built a strong personal network throughout the LEGO organization. This network had to be especially strong within the plants, where a good relationship with management would be a necessity and an even better relationship with the employees a vital ingredient for success.

Burke knew well that the previous track record of the LEGO Group when it came to developing a job training organization within the company had been less than successful. This was not the first time that the company had attempted it, and the two most recent attempts lacked, among other things, the necessary support from the local management. This meant that the needed resources were not made available, the sites were unable to agree on actual assignments, and an internal structure for the training organization was not developed or tested. Moreover, the lack of a true global steering group made the task ultimately impossible.

So, who could steer this important project safely and successfully to shore? At this time, there was a leader in the LEGO Operations Lean Office who had for a long time been addressing the idea of a learning center that could handle both HR and Lean courses for employees. He had understood that there was a great need for developing a learning culture, although few people at this time had seen the connection between Lean and a learning culture. This leader was selected to head the soon-to-be-established Global Learning Center. It was also decided at this time to organizationally anchor the Learning Center in the HR function in line with operational HR activities. The reason for this was to ensure coherence with other organizational development activities within the HR community and to ensure that the full focus would be on building workforce capabilities and knowledge transfer.

The next step was to find some people to staff the new function. After some initial negotiations, John Vellema (one of the coauthors of this book) was selected as the first employee of the new department and was recruited as the developer and global project manager. Previously, he had been part of the earlier training development and was at this point the project manager for the startup of mold manufacturing at the new production site in Monterrey, Mexico.

The first order of business was to make a current-state analysis of how training was performed at that time and what results were obtained from this training practice. The analysis showed that we were missing a systematic organization of processes and documentation that would support the knowledge and skills transfer from the home plant in Denmark to the local sites. Moreover, there were no local functions in place that could maintain and develop future LEGO Operations knowledge and skills through a sharing process. In Chapter 2, the current-state analyses are described in more detail.

One month later, our Lean leader arrived to head the new Global Learning Center. He and Burke worked hard to initiate the best possible start for the department. One of the critical issues, and also the most talked about, was where to start with preliminary meetings that would define the objectives for the Global Learning Center and the future role for the up-and-coming Local Learning Centers that would be placed within each LEGO Operations site. Would there actually be a need for these new departments, reporting to local HR directors? Or, would it be better to see the Local Learning Centers as "consultants" until the job training organization was in place? These kinds of questions were looked at repeatedly. We can say now, in hindsight, that it was a good decision ultimately to implement these Local Learning Centers as actual departments in each local plant. They now support the future development of local and global learning tools and ensure that the global knowledge-sharing processes are more effective between the sites.

It quickly became evident that there was a tremendous amount of details to be worked out. But, this did not change the fact that the main focus remained: How could we share production knowledge and skills more effectively across borders? We used several months to make a blueprint for the organizational structure of the Learning Center and how we wanted it to develop over the next few years. It was essential to create the right understanding of the processes at hand; we therefore spent a lot of time talking openly about past experiences and what felt right and what did not. One of the first things we could all agree on was that the new system had to be different from the previous unsuccessful systems. Furthermore, it had to eliminate the image of a system that was built from a Danish perspective and re-create it as a global process in which all could partake.

The main focus, then, of the initial discussions revolved around how to engage globally. This is why we decided, from the start, to have meetings with members from all levels throughout the global production sites—the needs analysis we referred to at the beginning of this chapter. In this way, we would gain much useful input and sow the seeds of interest and motivation among the employees by putting them in the loop. Of course, by obtaining feedback from a broader variety of people, it would be easier to create a system that would be effective and could be implemented on a global scale.

This turned out to be one of the most beneficial pieces during this prephase. By interviewing everyone—from vice presidents, production leaders, trainers, and employees in Denmark, Hungary, Mexico, and the Czech Republic—we received the input of all the stakeholders. This resulted in us becoming the "talk of the town." Actually, what was talked about was not us; the idea of a new Global Learning Center and the fact that there was a new global job training concept under development kept people talking. All of a sudden, everyone wanted to share their thoughts with us. Because we had visited and interviewed employees at all levels in all locations, growing optimism and confidence emerged within the organization.

It seemed to us that what really made all the difference in the end was the focus on the hourly paid employees, as opposed to focusing training solely on salaried employees and failing at this. This made our approach much more interesting to the employees, and they started to see this as a new beginning, believing that this might actually amount to something beneficial to them.

As we have stated repeatedly, one of the big issues for us to handle was the sharing of knowledge across borders. This is why, after 4 months, following the preliminary assessments, we brought on board a specialist from HR. Gitte Jakobsen (one of the coauthors of this book) was selected for this as she had great experience and expertise in learning processes, process optimization, and process knowledge and skill that she obtained through her work with LEGO production in Denmark and the Czech Republic. She was employed as HR manager in the Global Learning Center; her role was to help develop the new learning and development concepts and lead the Learning Center portion of the project.

This combination of unique skills in the Global Learning Center made it the ideal team for driving and developing the Global LEGO Training Organization. Both of the managers had strong interests within the area of organizational development and change, linking their work to the LEGO Lean journey as well as giving them a strong footing within project management. The mix of expertise in both HR and Lean, working in harmony, would prove to be the ideal combination of forces that would propel the effort to a strong start and a successful conclusion.

Chapter 2

Preparing for the Global Pilot Project

Introduction

With an enormous task at hand, structure and preparation were vital. We had four major manufacturing centers: the original plant in Billund, in the central part of Jutland, Denmark; the plant in Nyíregyháza, in the eastern part of Hungary, which had been run by a different manufacturer; a brand-new plant in Monterrey, Mexico; and a packaging plant in Kladno, Czech Republic, near Prague, which did assembly but no molding of parts. Because the project would include and integrate all of these locations, along with all the complexities of spanning their unique languages, cultures, and circumstances, it was pretty much a given from the initial concept stage that we would prepare and run a pilot project of our new organization to prove that it would work. Most of all, by showing the entire organization the power of a new approach to training and the results it could achieve, this would create a springboard effect that would launch the new structure to a strong and sustainable start.

We spent 8 months in total solely on the preparation of the global pilot project, and not one hour was wasted. Preparing for the pilot project was a critical and delicate process. We knew that we had to basically "padlock" all decisions once they were agreed on by all four plants. This meant that when a decision was made, it could not be changed afterward. We did this to ensure full commitment from all participants and thereby increase the possibility for success.

Some of the harder issues to agree on during this process were the number of hours to spend on the project and the project's success criteria. In addition to the decision padlock, before initiating the global pilot project we wanted, as a minimum, to have the collaboration and approval from the general manager, the Lean directors, and the HR directors from all four sites. Here, we describe how we organized the pilot project, determined its objectives, and created the key players and functions that would steer its implementation. The purpose of this chapter is to give a general overview of how the project was conceived. In Section II of this book, we go into the details of how the pilot project was actually run and provide a firsthand look at the challenges and rewards we reaped from the process. Then, in Section III, a more detailed description and analysis of the entire Global LEGO Training Organization is given, but the elements of this new system began to form even in the early planning stages. So, we refer to many different new positions and teams that began to form and were tested during the pilot, and more detail is provided about these in Chapters 6 and 7.

Pilot Project Team

The possibility for success of any project lies within the composition of the project team; this is why the task of appointing team members was not taken lightly. Carefully selecting members to make up the team, from the project sponsor to the project members to the project's stakeholders, was an immense task for a global organization like the LEGO Group. Finding just the right combination of expertise, enthusiasm, and leadership in each of the different entities who could then guide the project forward on a global scale took both time and diligence to pursue. But, these were the very people who would ensure the ultimate success of the project, so we had to make sure we had just the right people on board.

Sponsor

The project sponsor was the person who would have overall accountability for the project and ensure that the business need was valid and correctly prioritized. Deciding on the sponsor for our overall pilot project came naturally. As Stephen Burke was responsible for human resources (HR) in operations and, through his position there, could help on a strategic level with access to the right network, he was selected as the best person for this important position. Before Burke took on this task, he had previously performed extensive stakeholder management at various strategic levels. His work in these areas would ensure that the project team could gain access to all the right stakeholders to obtain and maintain their support in making the new Global LEGO Training Organization a success. This proved to be a wise decision both throughout the pilot project and certainly during the following global rollout.

Steering Team

Assembling the right steering team was the essential first step. This would be the team, not unlike a company's board of directors, that would oversee the work of the project team and ensure that all of its members were fulfilling the roles for which they were elected. They would also control and monitor both the funding and the progress of the project and its objectives. We knew already that the new training organization would have to be built to embody a globalized mind-set. This had to be done to meet the expansion needs of LEGO Operations. So, the steering team had to be built to reflect this reality.

We were fortunate enough already to have all of the general managers from the molding plants on board with the project; they, of course, would be members of the steering team. This meant involvement of factories in three different countries—Denmark, Hungary, and Mexico—which equaled three different sets of cultures and three separate mind-sets. In addition to this, the heads of molding design and implementation and the Global Learning Center were part of the steering team. It was determined that the general manager for molding in Mexico would be team chair, and he was, by far, the biggest supporter of the project. He had previously in his career started plants for the LEGO Group in Brazil and Korea. There, he had learned the hard way the importance of training when building a new site and was therefore enthusiastic to improve the knowledge transfer process both locally and globally.

The steering team would have the responsibility of making sure the project had all of the support and resources needed on a continuous basis throughout the duration of the project. This was no simple task. We knew well from previous attempts at creating training programs how easy it was for these efforts to "die on the vine" as new trainers, however motivated they may have been to succeed, failed to obtain adequate support from management. So, it was of utmost importance that we had key leaders in place and motivated to support and guide the project.

Project Manager

We needed to choose the right project manager, someone who possessed a unique set of skills that could guide an international project of this scale and content. It had to be someone with a good understanding of the business, excellent stakeholder management skills, and strong relations with all of the plants around the globe. The project would demand powerful global people skills so that the person could navigate successfully all levels of the organization on an international scale. As project manager, he or she would have to be able to see eye to eye with operators and, at the same time, communicate effectively with the steering team and stakeholders in their business network. John Vellema was selected as the overall project manager of the pilot and to lead the operations side of the project, and Gitte Jakobsen was to lead the Learning Center side of the project. (These two "tracks" are explained in the following section on the pilot project team.)

Key Stakeholders

In addition to the steering team, we wanted a close collaboration with key stakeholders, such as HR directors and Lean directors reporting to the general managers in molding. The benefit of having these individuals closely connected to the project was that this endeavor to build a global training structure was viewed by the organization as a gray area between these two departments, HR and Lean. In other words, both entities had an interest in the work that was being undertaken and a stake in the results we were trying to obtain with the project. We later found out that other companies had already bridged this gap in an area they called "Lean HR"; today, it is considered one of the keys for succeeding on a Lean journey: the human side of Lean. This turned out to be a good learning experience for us as we could understand firsthand what is meant in Lean by the oft-repeated, but rarely understood, adage of "respect for people."

It proved to be an important tactic that we included these two departments in the project as we worked across both of their areas. This was, however, not as easy as it may sound. Each of these two disciplines had their own priorities and vision of how training could or should be done, and there was a considerable amount of reluctance and skepticism toward the project. But, with patience and time, we were able to have them involved in the creation of the Global LEGO Training Organization, and we were able to use their input to fine-tune our own plans and strategy. As the results started piling up around the pilot project, most of these individuals became convinced in what we were trying to do and got on our side.

The Pilot Project Team

The pilot project was to consist of two separate tracks: a Learning Center Track and an Operations Track. Because the content of this work was training, we knew that we would have to have specialists in the field of adult learning and education. This would be the Learning Center Track. At the same time, we also knew that if we based our efforts solely on theory and not practical application, we would end up with smart systems that would, in the end, never amount to much on the actual production floor. So, we were determined to have equal representation by the practitioners, the workers, who would ultimately be responsible for doing the jobs. This would be the Operations Track. Together, the two tracks would constitute the pilot project team.

We appointed a track leader for each of the two directions; the main objective of these leaders would be to ensure high levels of academic and practical content while motivating the teams to achieve steady progress. In addition, an extra resource, a technical writer, was added to the Operations Track to help facilitate development of training materials and to assist in managing information technology (IT) solutions, such as developing a SharePoint page for the Job Breakdowns, which would come later with the implementation of the Training Within Industry (TWI) Job Instruction method.

Learning Center Track

It was decided to add a Learning Center consultant, and by "consultant," we mean a full-time LEGO employee who served as an advisor within the organization, from each of the three molding plants and these people would constitute the Learning Center Team. The Learning Center function was a new department in LEGO Operations, so all the consultants we needed to support the pilot project had to be recruited and trained in the local Learning Center functions. This was done to ensure that the internal concepts developed in the project would be correctly presented at all sites, a challenging task that was made even more complex by the need to have them translated properly into the local languages. The process of implementing and maintaining new concepts into a well-established organization is difficult at best, and we needed people who had the skills and expertise to do this well.

Some of the plants had already established internally, outside our project, a training consultant role, while others would have to employ a person for this specific task. The job description could best be described as a mix between an HR consultant, with special focus on organizational learning, development, and change, and a Lean consultant, with expertise in processes. The job assignments for the Learning Center consultants were broadly defined, from developing HR processes to selecting the best job trainers to facilitate the development of Job Breakdowns, a major tool of the TWI Job Instruction method, for training.

As a last-minute addition, it was decided to invite a Learning Center consultant from the plant in the Czech Republic to participate in the first workshop. The reason why this plant had not been included in the initial analysis phase was that its core ability lay within packaging, while the focus of the pilot project was to be only on molding. But, by including this plant in the project, we not only would be allowing one of their Learning Center consultants to experience the workshop, but also would benefit by seeing whether other areas within the LEGO Group could see the possibilities and advantages of the work we were doing. It would also help ease the process of a global rollout later if we had an "insider" connected to each of the manufacturing plants. After the first workshop was completed, the Czech Learning Center consultant announced that he would be quite pleased to participate in the rest of the workshops because he saw a great deal of relevance in the work we were doing (see his case study in Chapter 9).

Operations Track

The three key employees selected from production to become members of the project's Operations Team, one from each of the molding plants, had all previously worked in job training planning at their individual sites, and they all had a profound desire for the training function to expand and reach a new level of global sharing. The clear benefit of having these training veterans on the team was that they had already established a strong network within LEGO Operations and had experience working together. They also all had previous experience with training in molding.

Two of the three had vast knowledge and hands-on experience when it came to molding, having served as molding operators earlier in their careers. The third member, on the other hand, was new to the business and was a strong process engineer as opposed to a practitioner.

This combination of varying skills and characteristics, however, turned out to be just the thing we needed. When the Operations Team started developing material for the classroom teaching, it was evident to us that the two with the hands-on experience were great at going into the details of the jobs and seeing them from the trainers' point of view. They basically had a more practical approach. Our process engineer, on the other hand, was more focused on the overall process rather than on the details of the jobs themselves, and she became the force behind accomplishing the overall work.

These three individuals became our functional area master trainers (FAMTs), and these individuals might be identified in other organizations as either training coordinators or training managers. Together with the newly appointed FAMTs, we started the selection of global job trainers (GJTs), what other organizations might refer to as training supervisors or training staff (see Chapter 6 to learn more about these positions and the roles they played in the overall organization). From the start, we determined some common criteria for the position of GJT: All had to be incredibly good as setters (the person who does the mold changeovers), to have shown unique capabilities in training coworkers, and to have proven ability to contribute and participate on a global team. On the "softer" side of the personal profile, we were also looking for strong individuals who local coworkers would view as informal leaders because this would help later with the postproject implementation when they went back to their own working areas.

Another thing with the GJT is that we wanted two employees from each factory to hold this position because, when it came to making the Job Breakdowns, we saw a clear benefit of having two people from the same factory instead of just one. This was not only from a perspective of "two minds think better than one," but also the knowledge that the GJTs might be placed under a lot of pressure if they were held responsible for this task by themselves, so from a support point of view it would be better for them to work with a partner. They were the ones who would have to implement and enforce the Job Breakdowns when they went back to work in their home plants, so it was essential that they have the resources to accomplish this.

It took a great deal of time to structure the resources and assemble this team in all three molding plant locations. But, it paid off in the end because of the decreased time it later took for us to begin delivering excellent training material.

This brings us to the last important skill the GJT needed to possess. Because this was a global project involving several different plants in different countries, it was essential for us to have a common language for communication. The last required skill was for at least one individual of the two-person teams at each site to be fluent in spoken English. We were lucky. At least one of the GJTs we had already selected from each of the plants was able to speak English. Moreover, it turned out that the ones who were not fluent in English, or were lacking in the skill, picked it up quickly to enter the discussions and have their say during the Job Breakdown sessions. We saw a strong team spirit develop, with the GJTs helping each other and supporting the team both in developing the Job Breakdowns and in working globally and learning English.

Pilot Project Description

Creating total agreement on a project description between all of the general managers, HR directors, and Lean directors across the various borders proved to be extremely demanding and required an enormous amount of time in negotiation and stakeholder management. The real issue was to secure a common engagement to the mission at hand that everyone could agree to pursue. Without this common engagement, it would be impossible to prove that going from unstructured training to a more professional global training organization using the TWI Job Instruction method and Job Breakdown format would indeed improve the global knowledge-sharing process.

Background

As we learned in Chapter 1, the "burning platform" for the pilot project was the LEGO Group's growth strategy, to which the steering team had some clearly defined demands—two to be exact. Both of these demands supported the global growth strategy.

1. Increased efficiency in ramping up the workforce.

   The blueprints for a new superfactory in Mexico had already been created, and the plant was to be built in the near future. We also knew that future expansion in Hungary was being talked about as well, and even expansion to Asia was on the drawing board.

With a growth strategy as aggressive as this, we knew we had to think about training in a completely new way. It needed to be much more efficient. It is a big endeavor to build a new factory, but it is a much more daunting task to have to train so many new people in so little time. In addition, we wanted to build independent scalable plants where in the future local staff could train new coworkers on their own.

2. Increased efficiency in daily operations while maintaining compliance with safety and quality.

   This point focused on stabilizing production processes and, through this effort, minimizing the amount of variation between countries, plants, shifts, and individual employees. The main focus of this second demand was safety and quality, two focal points that would later become the foundation and the core for the Job Breakdown sessions that determined the content of the job training. This was the best possible direction for the development of the project because it created buy-in later when implementing the job training.

Besides these two demands, there was also a third demand: a demand for a change in culture. This culture change was actually a necessity for the first two demands to succeed. We saw an evolving global footprint for the supply chain that would drive a need for our company's employees to possess both a global and a local mind-set. It is always a challenge to go from a one-way-driven knowledge flow to global knowledge sharing. To do so, we had to change the mind-set from, "Not invented here," to a more global mind-set of, "The more we share, the more we get," and "Copy with pride."

When you take over a plant or build a new one, there will always be a critical phase at the beginning when a common culture is created. It goes without saying, but if this phase of the changeover is not approached with care and skill, things can quickly spin out of control and create an unproductive and negative culture—just the opposite of what we want and need. This is why we took the aspect of good global culture into consideration from the start and made sure it was handled in the correct manner.

By making sure that everyone had their say and received genuine understanding and appreciation, we were able to create a higher level of unity through effective dialogue.

Objective

The primary objectives for the project had now been narrowed down and clarified for us. We had come to the following conclusion for our primary objective:

> Improving local and global alignment of processes and knowledge sharing within and among functions in LEGO Operations sites by
> Developing a global job training organization that can bring on board new employees and improve the competence level of our current employees by improving the way we design, execute, and evaluate our training at all sites.

Needs Analysis

As a preanalysis to starting the pilot, we needed to obtain an in-depth feel for the current situation at the molding plants, what we called our needs analysis. Meetings with the plants in Mexico, Hungary, and Denmark (at this point, the plant in the Czech Republic was not in the planning; this plant came in later, after the first workshop) were set up to expand our understanding of the current training organization process. At all three sites, we spoke with workforce operators, trainers, first-line leadership, middle management, support areas, and top management. This supplied us with a clear image of the situation and their ideas for an improved global training organization. All this input was used in designing the new global training organization; without these discussions and stakeholder involvement, the design phase would have been difficult and the chance of success for the pilot low.

Visit to LEGO Operations Mexico

The factory in Monterrey, Mexico (LEGO Operations Mexico, LOM), was as new as they get, only half a year old, when we started and they had just begun undertaking the task of training. The factory was built in this region to have production located right between North and South America, so strategically the location was ideal. Production was to expand from a current level of 50 molding machines to an expected 850 units or more. The implementation of new machinery was to be done in clusters, thereby leaving room for the necessary training of new employees.

Production at this point was already being pushed, so trainers from the Monterrey plant had traveled to Denmark for a head start. The training sessions in Denmark were carried out based on prior experience, and the head of training in Billund created a training plan based on the Danish methods. It should be mentioned that due to a lack of time, the preparations were not what they should have been. The Danish trainers in charge of producing the new training material were not sufficiently prepared to go over basic content with their new colleagues. Above all, no one was prepared for the lack of cultural understanding between the two groups, and because of the language differences, after they were finished there was no real hands-on experience for them to lean on or to refer to. So, it was a big challenge to anchor the knowledge and skills in the Mexican organization and to ensure the self-sustainability of the site.

There were several reasons for the difficulties faced in the initial Mexico training program. Primarily, it was the number of trainers involved with training the employees and the methods they were teaching. As it happened, different methods and processes were taught by different trainers; this was creating misunderstandings among the workforce. Moreover, Mexican employees were more loyal to written manuals as well as standard operating procedures (SOPs) than to the know-how given them by trainers. So, in addition to receiving different information about the same procedures, the instructions they were receiving also deviated from the original SOPs, compounding the confusion.

On a higher, perhaps more profound, level, depending solely on "experts" from the home country and not showing the plant the respect deserved by turning over responsibility for training to a locally established group seemed to be just as big a factor in the challenges faced in the initial round of training as any of the content delivered. Physical presence on site and experiencing the day-to-day operations and life in the plant are fundamental to creating the best solution for a training organization. Assuming that trainers from or in Denmark would always have the best information and techniques did not allow local Mexican workers to take full responsibility for their methods, and this lack of ownership was a roadblock to resolving the issues of inconsistent work instructions.

The Hungarian Phone Calls

The LEGO plant in Hungary had been taken over from another manufacturer a half year before the project started. The head of job training in the molding area at the Hungary plant was an educated craftsman from Denmark who had little HR experience and did not pursue a strong Lean approach in operations. Nonetheless, during phone meetings with him, he pointed out several interesting and crucial issues in the operations part of the production in Hungary. First, he highlighted that most of the training material he received from Denmark was of inconsistent quality, so he had chosen to create and produce his own training material. He furthermore noted that a training program was in high demand. To standardize the work processes even further at the Hungarian plant, he felt they would need just such a plan which would then allow them to calculate the amount of time they would need to allot for training.

As a last thing, and definitely an important point, he made it clear that a cultural program was needed as well. Even with little expertise in HR, he could understand and grasp the clear benefits of closing the cultural gaps when creating a global training organization. In his case, he was experiencing not only the cultural friction of trying to implement Danish training materials in Hungary but also the differences in the business and work culture while changing from the former owner's system to the LEGO model.

Discussions in Denmark

John Vellema had earlier been part of the initial training sessions which were then held at the Mexico site with the new employees there. So, he had a good understanding of the situation as well as a strong relationship with the Danish trainers who went to Mexico to train. The trust and bond he had with the Danish stakeholders in molding gave him the opportunity to speak freely and receive honest input from them. He quickly found that the Danish trainers were frustrated because they knew they were able to perform and deliver a much more solid product, but because of time pressure, they were not able to prepare properly for the training itself or follow up with the employees after the training. Moreover, the trainers made it clear that they wanted stronger communication among themselves because, in fact, they had encountered instances when their training had differed from other trainers. This created issues of trust toward the trainers.

More than one trainer had expressed frustration when they were not given the opportunity to train any local trainers, who could then continue with the training of the local workforce after the Danish trainers left the foreign plants. This input further supported our idea of creating a global job training organization that was locally anchored at all sites. The Danish trainers and training coordinators had great experience training in the LEGO molding areas, and their input helped us considerably.

Summary of Local Visits

After visiting the local sites, we went through all of our research, all the interviews, and all of our individual impressions. We did this to obtain a clear-cut picture of the current situation and to understand just how much effort it would take for the new Global LEGO Training Organization to become a success.

We knew from the start that it was essential that adequate resources be given to a project of this magnitude. We also understood from the interviews with the Danish trainers that a lack of time and resources had greatly hindered their efforts. But, perhaps even more importantly, we realized that we needed to have the full cooperation and willingness to change from the actual people involved. This was one of the first discoveries we came across during our visits. There was a genuine excitement and positive attitude toward this change and improvement, and this came from all the LEGO Group employees we came across, from upper management down to the shop floor. Both new as well as experienced workers were motivated to change how we practiced training. This was a real win for us as we felt we had support from all levels, creating real motivation on our part for the project.

Gitte Jakobsen's favorite adage, "The degree of preparation is equal to the degree of success," was put into play here. What we learned from the analysis was the necessity of having a consistent organizational structure at each plant and a common language to improve the knowledge-sharing process on a practical level. As we saw during the Mexico visits, it was clear that we needed everyone involved to understand that there was no such thing as the "Danish way," there was only the "LEGO way." Furthermore, everyone needed to pitch in to create the best possible training environment. This was essential to the process, not only when it came to orchestrating the pilot project but also as one of the fundamentals for involving workers and obtaining their essential tacit know-how.

Selecting the Pilot Case

We wanted to ensure our pilot case had a visible and direct connection to the stated organizational goals. It had to be a key issue that needed to be solved. It had to have high visibility, be limited in scope, and be measurable according to current KPIs (key performance indicators). All these demands for the pilot case were analyzed and debated in collaboration with the steering team and project stakeholders. Since molding is the core production competency for the LEGO Group, the choice was obvious to test it in this discipline. We did, however, have some concerns because the complexity and scale of the jobs carried out in this specific area were immense.

There were various aspects of the molding operation that made it the perfect pilot project for us. First, the newly opened factory in Mexico was looking at a large expansion, as noted previously, and an expansion in Hungary was also a possibility. It was evident that the molding area contained more complex tasks than any of the other possibilities. The molding area dealt with precision work; in addition, it was performed on a number of different shifts carried out by craftsmen workers. All these factors, and more, combined to make it a clear and "easy" decision.

We wanted a pilot case where we could test the TWI Job Instruction skill, which we had selected as our method for doing the instruction, in a realistic environment and, as an added bonus, in an area that was not "just" the easiest area possible. After all, the whole purpose of piloting a new method is to be able to show that the method can take on a complex task. It will also help when transferring the method to other different areas because you will be able to show the range of assignments to which the new method can be applied, from simple tasks to the most complicated procedures. On the other hand, we had to be careful not to take on too difficult a task because the method was new and something everyone was still learning to use. In other words, it had to be "doable." This again was one of the main reasons for our choice of molding as our pilot case. We found the molding process would be achievable for us to demonstrate the method yet complex enough to convince top management of the need for designated training.

After choosing the overall area for the pilot case, we approached the three plant directors to obtain a clearer picture of which tasks in the molding process they thought should be used as pilot jobs. They suggested that we focus the pilot on the mold changeover process. When we approached the steering team with a similar request to help us find the most suitable assignment for the pilot project, it was obvious to them as well that we should choose the changeover process of a casting mold. We knew, though, that by taking on such a loosely defined assignment, things could quickly spin out of control, so we asked them to narrow it down to a specific, small, yet essential assignment. We walked away from the meeting with the pilot project defined as, "Mold change for VC Engel 60 ton molding machine, with a three-plate mold, working with ABS-type plastic." This assignment might have seemed an easy choice at the moment, but it would turn out to be more difficult than first anticipated.

Chapter 3

Setting the Foundations for Moving Forward

Introduction

From the beginning of the project, there had been a clear understanding between the steering team and all the relevant stakeholders that they wanted the plants to work across borders, and that a strengthened global team spirit should characterize the shape of the pilot project. One of the paramount goals, then, would be to strengthen unity across nationalities and cultures. Based on this clear objective, it was decided that the project model should be molded around collaborative workshops at which, together with their counterparts from all the other plants, the Learning Center consultants would develop the internal concepts, the global job trainers (GJTs) would create and explore the process of making Job Breakdowns for training, and the functional area master trainers (FAMTs) would develop the trainer recruitment plans and the classroom training material for mold setters. The word *together* was the key concept, and team members from all of the plants would be equally involved with equal voices. This created a unique work environment and a profound wish to demonstrate that our motto, "The more we share, the more we get," was, in fact, more than just a platitude, but a concept that could be acted on and would create the future culture of the Global LEGO Training Organization.

This team spirit could be seen manifesting itself during the pilot project in the various development phases of the tools and methodologies the team was creating. During these decision-making periods, it became clear just how global their mind-sets had become and what they were willing to sacrifice to reach their common goals. They divided assignments among themselves and set meetings at workshops around the world to share, develop, and narrow down the essential details of whatever they were working on in a way that generated the maximum benefit for all in the time given. This was done collaboratively to obtain the best-possible result for the course of action at hand. This was clearly a different path than what the LEGO Group had experienced in the past with cross-cultural efforts and activities, for which the Danish participants would simply promote the headquarters' doctrine, and local participants would struggle to maintain their own positions and integrity.

From efforts such as these, we learned, after just a few workshops, that the team always focused intently on maintaining a consistent group direction no matter the source of the ideas or suggestions. The paradigm that good ideas could only be generated from the headquarters in Denmark had been rendered obsolete. This created an underlying and fundamental trust between the sites, which still exists today in the global job training organization. This trust encouraged all of the members to pitch in on the development of the best training and trainer development concepts and to create them with a sense of unity. Having to collaborate across borders and cultures also meant starting out developing the content in English while making sure it would be relevant and effective whether it was implemented in Mexico, Hungary, the Czech Republic, or Denmark.

This way of working together became such a big part of the mentality of both teams (Learning Center team and Operations team) that when they met during workshops or corresponded internationally, they developed a common terminology that facilitated the communication between all the members. This common terminology contributed to the success of sharing knowledge and new ideas, from local to global, and that made the entire effort much easier and more effective. It furthermore contributed to a more open and positive work environment as well as a more independent one. This might seem contradictory, but in a standardized work situation where contributors no longer need so much facilitation or direction as before, they become more *independent* because of the standardization.

The global workshop model we pursued, with the whole team meeting at different sites at different intervals and having assigned "homework" during these intervals, was inspired by a networking meeting we had had previously with a large Danish pharmaceuticals company. That company had initiated

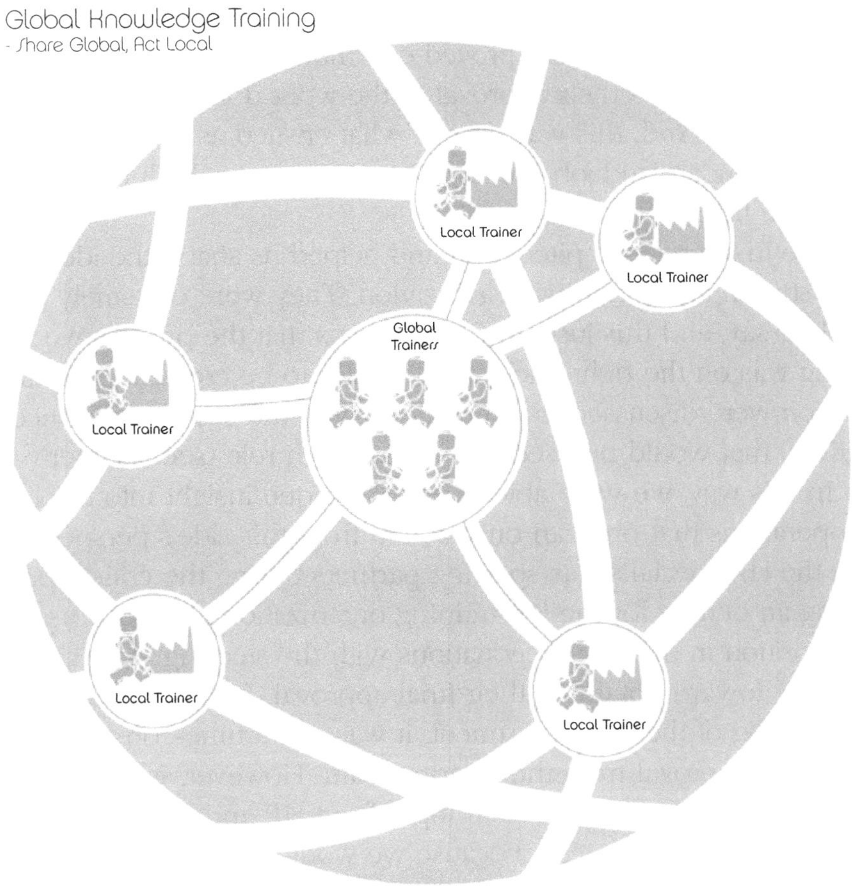

Figure 3.1 Global knowledge training: share global, act local.

a similar global standardization project and spoke highly about this model and the great global teamwork it created. We especially liked the idea of *sharing globally* and *acting locally* because it is a great way of creating global team spirit (see Figure 3.1). By reading this chapter, we hope you obtain a sense of how effective it really was and how important it was to grow the team through face-to-face meetings from the start.

First Draft of the Training Organization

Fortunately, the initial meeting of our team coincidentally occurred at the same time and place as the semiannual HR Face-to-Face Meeting, which is a regular gathering of human resources (HR) directors from all LEGO sites. This turned out to be the best thing that could have happened at this early stage.

Having the opportunity to present our ideas at a meeting with this amount of expertise gathered in one room proved enormously beneficial and the only thing we hoped for was their approval of the work done to that point—and loads of feedback. And, this was exactly what ensued as it became evident that a global standardized job training organization was high on all of the HR partners' wish lists.

Everyone in attendance pitched in and helped us shape the ideal structure for just such a job training organization. They were extremely eager to have their say, and this just confirmed for us that the process we were developing was on the right track. All the roles to be created under the new organization were discussed in detail along with the key performance indicators (KPIs) that would be used to evaluate each role (see Chapter 6 for details). In this way, we were able to gain a unique insight into the human side of operations that one can only obtain from this side's perspective.

Using the HR specialists as sparring partners during the critical process of developing an outline for the job training organization also gave us a much stronger position in the final negotiations with the steering team and helped ease the way toward obtaining their final approval. Even though we now had the backing of the HR department, it was still a time-consuming process to obtain final approval from the steering team. However, we understood that taking the time up front to gain input from HR and other areas would save us time in the longer term because we would not find ourselves caught in a situation debating specifics later in the process when we were actually testing the methods. By clearly defining the process up front, we could approach the project with confidence knowing that the various stakeholders were all on board.

Choosing the Training Within Industry Program: Why Job Instruction?

During this preparation phase, it became clear to us that we needed to select a defined training methodology as well as a concrete means of describing and documenting the job knowledge and skills. As we found out, our current trainers did not have strong tools to support them in preparing, executing, or evaluating the training they were performing. When we realized just how much professional and practical knowledge and skills the trainers possessed, we were shocked to find, in spite of this, so many instances where they were struggling to train successfully because of the lack of a consistent method. John Vellema remembered that, during the time he spent at the LEGO Concept Factory, one of the production leaders had approached him and handed him a Danish Training Within Industry (TWI) Job Instruction Card, the small pocket-size card you receive during the class that spells out the four-step Job Instruction (JI) method as well as several points needed for preparing to train. This small gesture reminded him of a report he had made on TWI for a Lean project during his earlier studies as a student, when he worked for a wood-processing plant in Brazil. It was actually this work, and the reminder of it he received later with the card, that spawned the idea of using the TWI JI training method for the global training project.

The reason for selecting TWI JI as a training method was evident: It had clear rules for creating training materials, the teaching method was extremely structured, and it would be able to give our trainers the necessary tools to achieve success. Before we decided on TWI as the defined method, though, we studied several business cases for which success was achieved using TWI. In addition to these, we read up on TWI and its definitive link to Lean through the books *The TWI Workbook* (P. Graupp and R.J. Wrona, Productivity Press, 2006) and *Toyota Talent* (J. Liker and D. Meier, McGraw-Hill, 2007) and a number of articles. So, this was not an overnight decision.

The only real worry we had with the TWI training method was whether it would be able to deliver on and live up to the newer, more current training theories. In addition, there was a growing concern in our group whether trainers would really have the time to prepare, perform, and follow up on the TWI training they gave. Without time to do the training properly, we would not be able to prove if this method really worked. We did, however, appreciate greatly the simplicity of the JI method and the fact that it was tested over many decades of time. Moreover, it was developed as a train-the-trainer style of education, which ensured all trainers received the same instruction and therefore would train jobs to the same standards on the shop floor. In the end, the pros outweighed the cons, and we chose to go with the TWI JI program. (It is outside the scope of this book to explain the content of the TWI JI method, the history of the program, and why it has been, for over seven decades, such a powerful training tool. We recommend the two books mentioned, which will serve as excellent companions to this one, for your reading and study.)

Our first phone call with the TWI Institute in the United States basically revolved around the cost of implementing the TWI JI program, which was a challenge for us since it had to meet our budget.

By that time, we had already burned through many resources in the form of man-hours and extensive travel activity, so our budget was tight. But, it was time for us to make a final decision and place the order. We had determined by ourselves that, due to the extensive research we had already done on the program that we were ready to move forward and begin training in-house trainers on our own. In her first telephone dialogue with Robert Wrona, executive director of the TWI Institute, Gitte Jakobsen tried to convince him that we were ready to go directly for the institute's 40-hour train-the-trainer program based on our thorough prework and study of the method. He did not, however, share this same optimistic perspective.

After having introduced Wrona to the project goals, the project plans, and our thoughts on the TWI program, we tried to justify our motives for "skipping" directly to the 40-hour program. But, Wrona explained to us that this would not be possible as the institute would not let anyone, under any circumstances, take the 40-hour trainer development program unless they had previously passed the 10-hour basic JI class. This meant, he told us to our great dismay, that we would need to sign on first for the 10-hour JI program and, in addition to this, dedicate another 4 to 6 months practicing the TWI JI method *before* even starting the 40-hour trainer development program. Only then, he explained, would you understand the correct usage of the method and be able to spread it through your organization. At the end of the meeting, he simply stated:

> Either you have wasted a huge amount of time or you are really well prepared for the change. If I am right, you are going to succeed with the global pilot project.

This was the tipping point for us. We now understood that, having spent so much effort on this, it would be a mistake not to start with the 10-hour program and to do it the right way right from the start. It also told us that Wrona was not just trying to sell us a program but actually respected the huge effort we had put into the project and did not want us to let it go to waste. In addition to this, he explained the importance of the TWI Job Relations (JR) course, the part of TWI that developed people skills, which we rejected outright, reasoning that we in the LEGO Group had a great company culture in which there was a huge amount of trust between the employees and their leaders. He also recommended we do an additional 3-day TWI JI Follow-Up Coaching course.

We chose to sign up only for the 10-hour TWI JI course and had to reject Wrona's advice about both JR and the follow-up coaching. Because our budget was really tight at that point, our plan was to test the JI program first and let it prove its worth. If JI produced good results, we reasoned, we could try to gain more resources later for other needed programs like these.

The Link between Job Instruction and Lean

If you want to reduce the degree of variation within a defined area, whether for reasons of safety, quality, delivery, or all of these, the procedure many would and should turn to is creating and maintaining standard work. And, we know from our study of Lean that the foundation of any continuous improvement system is standardized work. To achieve this needed standardization, though, there are no quick fixes, and you will have to undergo a long and hard process of change. At the LEGO Group, the demand for a solution to the global training need was urgent. This created many challenges with the huge transformation of going from unstructured training to structured training. But, it was absolutely vital for us to meet this change head on and to make a difference in how we trained our people to perform standardized operations.

Here, we realized the importance of a more nuanced approach to a task that demands such a big cultural change. If you are unable to change the company's culture, then, ultimately, you will not achieve your long-term goals. Liker and Meier pointed this out clearly in *Toyota Talent* (p. xxii) when they described how purpose can define the results you obtain:

> It is essential, however, to place importance on the means rather than the end. If your motivation is simply to do a better job of training because it will reduce problems and help your company, your outcome will not likely match Toyota's. This is what we would call a "results orientation" rather than a "process orientation." Clarify your expectation up front and make sure you have the proper orientation. Do it because you genuinely care about people and are committed to the effort required to help them achieve their greatest potential; in the end, you will get the best results.

The best approach would be to set a clearly defined goal and follow it up by asking, "Do we have the organizational structure and the essential competencies to reach the goal?" Then, look at the "gap" between the current situation and the goal. In Chapter 2, we saw clearly from our needs analysis the distance separating our goal of consistent global training and the reality of how training was actually accomplished. Bearing in mind that we were dealing with human beings, we set out then to define and institute an organizational structure that created the essential competencies we would need to achieve our goals.

Achieving lasting change is not a quick fix but more like a lifelong process. When creating standardized work, you should always start the process by methodically building a strong foundation. The foundation for standardized work is developed though a stabilization phase during which effective job training will teach operators how to perform their jobs following a defined method so that each worker, at each station, on each shift, does it the same way. The main focus for our pilot project was *not*, in fact, to develop a foundation for standardized work but to share knowledge and experience between plants. But, what resulted from this effort was, indeed, to create a strong foundation for standardized methods; this was a huge added bonus. It became a means to an end, and once the Lean Office saw the possibilities for true standard work in the foundation we were building, they quickly became committed to the program.

A Strong Lean Foundation

In the midst of trying to see how our project fit into the overall Lean strategy, we were excited to learn that Toyota, in its long history of Lean development (better known as the Toyota Production System [TPS]), had relied on TWI JI as one of the strong pillars to becoming the world's most respected company in process improvement. Toyota began using TWI in the early 1950s, and it became the basis of implementing standard work, which became a benchmark to all of the other facets of the TPS. As Isao "Ike" Kato, who retired from Toyota after 35 years and is known internally in Toyota as the father of standardized work and *kaizen* courses, put it:

> I don't think you can do a good job of implementing standardized work or several other elements of TPS *without the JI skill set in place*. I have observed quite a few companies struggle with implementing standardized work, kaizen, and other items. Often the short

> term gains companies obtain fall away over time. One direct reason why is that no proper plan was ever put in place to train people to the new method and the JI technique provides the exact skill set required to do this work. *I can't see how standardized work can function without JI in place underneath to support it in the long run.* (emphasis added)*

We chose, then, to start the project model with a singular focus on *standardized work* because of the need for developing competencies within the organization. As you can see from the TWI/Lean integration model shown in Figure 3.2, JI creates stabilized methods that become the foundation for standardized work. JI, in turn, thrives in a stable working environment that is built on good job relations, safety, and good housekeeping. Once this stabilization begins to take shape, then, and only then, can we see what methods are working over time, based on quality, safety, and output data, and then lock in these methods as "standard work" that can form the basis for future continuous improvement efforts that will build on these standards. In this way, the TWI programs of JI, JR, Job Methods Improvement, and Job Safety integrally fit with the tools of Lean because, in effect, they were

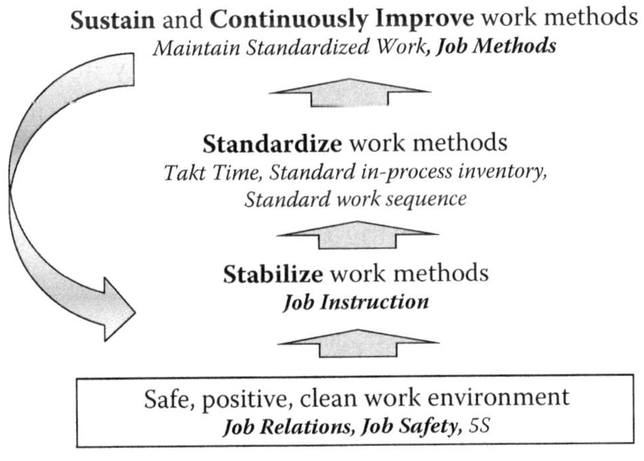

Figure 3.2 TWI/Lean integration. (From P. Graupp and R.J. Wrona. 2011. *Implementing TWI: Creating and Managing a Skills-Based Culture*. Boca Raton, FL: Taylor & Francis Group/CRC Press. With permission.)

---

* From *The Art of Lean*, http://artoflean.com/documents/pdfs/Mr_Kato_Interview_on_TWI_and_TPS.pdf. (p. 4, paragraph 3)

in place first while these Lean tools were being developed following TWI's introduction to Japan just after World War II.

As we alluded to at the beginning of this section, the next two issues to be investigated were

1. Understanding which competencies the different roles demanded to reach the objective
2. The size of the gap to close between actual competencies and needed competencies

Both of these are defined by looking at the current organizational structure and the competencies existing in the current state. But, to make a correct evaluation, the evaluator must have deep insight into the true aptitudes and abilities of the people doing the work as well as a thorough understanding of the unique requirements of the processes. This would entail both a functional knowledge of the processes, similar to an engineer or technician, and a working understanding of human cognition and potential, similar to a psychologist or teacher.

To meet this demand, the Learning Center team, who would be in charge of making these evaluations, was a combination of employees carefully selected from both the Lean and HR departments. All of the members of the team had strong backgrounds that included experience within both areas, and we made sure that they actually had hands-on work experience. We saw this as an essential quality, a necessity really, to be able not only to understand both worlds but also to have an understanding of the link between the two fields. This is what makes a great Learning Center consultant, taking the best of both "schools" and merging them into a plan of collaboration and mutual support.

Not many people have this combination of skills and experience, and it is often an either-or situation. When they could not be found, we paired up one Lean employee with one HR employee to make sure both participants kept an open mind toward each other's area of expertise. Creating this team of Learning Center experts in our pilot project made the difficult task of understanding which competencies were needed and available more effective. After the project was approved, we established a close collaboration between both the HR and Lean departments to continue this support of the organization in such a positive fashion.

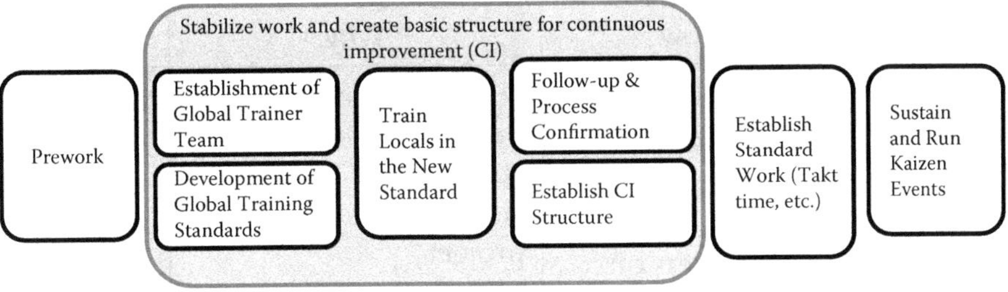

Figure 3.3 Standard work development model.

The Standardized Work Development Model

The model in Figure 3.3 captures our experience from the pilot project in molding and our thoughts on how to implement standardized work. Here, we give a high-level overview of the process from start to finish. In Section II, we go through the details and inner workings of the pilot project to see what it looked like from the practical aspect of actually carrying it out from week to week and month to month. Then, in Section III, we go into the details of the overall organizational structure of the new global training system, the details of the roles each of the participants would play, and the mechanisms that would make the entire system work smoothly and effectively.

Each step of the model matured the work process, the training organization, upper management, and the workforce. This model represents what we felt needed to be built to implement standardized work. Depending on the process and the department, the different phases might vary, and this would shorten or lengthen the time to go through each phase. But, the fundamental concepts and progression of development remained the same.

The following descriptions of each phase of the development process, with examples of its content, are more guidelines than thorough lists of step-by-step implementation points. The last two points in the model, "Establish Standardized Work" and "Run Kaizen Events" can only be reached after a stabilized work flow has been implemented.

Prework

The prework phase ensured that there was a good plan that had the commitment of all involved. Furthermore, the target here was to give the team a good start locally using the experiences of other areas that had already implemented the plan.

What	*How*	*Why*	*Support*
Development of A3* and TIP (Tactical Implementation Plan)	Call in relevant people Ensure consensus with all stakeholders on the A3 and TIP	To ensure the right buy-in from all stakeholders before starting the actual project	Learning Center
Selection of area to start	Analyze area to identify development needs (e.g., poor FTT [first time through] rate)	To start with an area that gives value, first go deep with one process and then wide in the area	Learning Center
Establishment of baseline measurements (e.g., FTT)	Observe production and provide questionnaires	To be able to track the improvement over time; focus is on knowledge in relation to safety and quality	Lean Office
Selection of the needed trainers (FAMT, GJT)	Use the Learning Center selection process	To recognize and develop potential trainers	Learning Center
Local kickoff	Give information about plan and concept	To create motivation and a clear picture of the plan	Learning Center and Lean Office
Introduce trainers to TWI JI	Provide the 10-hour TWI JI class	To be able to break down various types of jobs; to break down complicated jobs into teachable skills; and to improve employees' skills	Learning Center
English language training (where needed)	Attend English language training on a regular basis	To be able to communicate with trainers from other international sites	Learning Center
Follow-up of training	Provide quarterly for workers	To be able to measure progress	Learning Center

* The A3 Report is a Toyota-pioneered practice of getting the problem, the analysis, the corrective actions, and the action plan down on a single sheet of large (A3 size) paper, often with the use of graphics and illustrations.

Establishment of Global Trainer Team

This phase was to build a strong global trainer team; the better the foundation in this phase was, the better the training material development and knowledge flow would be later.

What	*How*	*Why/Target*	*Support*
Global workshop	All sites meet to form team, share knowledge, and develop global standards	To build the needed relationship and trust so a well-established foundation of knowledge sharing is in place	Learning Center
Global kickoff	Information and orientation regarding plan and scope	To provide common understanding of plan and scope	Learning Center
Awareness training and "working global" training	4-hour cultural awareness training for the global team	To build a better understanding of team culture and thereby prevent conflict from arising	Learning Center
Team objective, values, and rules	Team session where the team develops its own unique values and rules	To create a solid foundation for future work	Learning Center
Social event	When the team meets at a workshop site, host site prepares a social event	To develop strong relationships and team building	Learning Center

Development of Global Training Standards

To be able to move on to global continuous improvement at the end of the development cycle, foundations for the jobs needed to be in place. Through development of global training standards, all sites would be working in unison. This would provide a unified global language so that they can work together on these jobs at LiveMeetings (Internet conferences) in relation to continuous improvement and share coming standards in the relevant job areas.

What	*How*	*Why/Target*	*Support*
Development of global training materials	Delineation by FAMTs and GJTs of jobs to be performed by employees; breakdown of jobs into teachable units	To create the basis for standardized training of employees worldwide	Learning Center and Lean Office
Basic knowledge training	Development of classroom training materials related to basic and specialized knowledge		Learning Center
Skills training: (1) Standardized Work Chart (SWC); (2) Job Breakdown Sheet (JBS); (3) Work Element Sheet (WES)	Development of shop floor training materials related to basic and specialized knowledge; breakdown of jobs next to machine and involving relevant persons		Learning Center and Lean Office
Skills matrix	Development of overview of jobs (who can do which job) in the area by involving relevant persons		Learning Center and Lean Office
Training plan	Development of training plan for employees by involving relevant persons		Learning Center
Global workshops (only first standard)	Development of above materials in global workshops, LiveMeetings, and at local site		Learning Center
LiveMeetings	Development of above materials in global workshops, LiveMeetings, and at local site		Learning Center

Training Locals in the New Standard

To be able to train all employees in the new standard, we needed local job trainers (1 trainer to 10 employees); this would ensure that all employees would be trained by a trainer who was prepared and understood the nuances of adult learning. Most important, every person learning the job would learn it in the same way regardless of local language and culture.

What	*How*	*Why/Target*	*Support*
Local job trainers selected	Learning Center selection process	To recognize and develop potential trainers	Learning Center
Local job trainers introduced to TWI JI	10-hour TWI JI class	To be able to break down various types of jobs, to break down complicated jobs into teachable skills, and to improve the employees' skills	Learning Center
GJTs train local job trainers in new standard	Training plan	To be able to train local employees in new standard (continuous improvement starts here)	Learning Center and Lean Office
Local job trainers trained in LEGO training and learning skills	Classroom training by Local Learning Center	To obtain awareness, for example, of adult learning principles, knowledge of yourself, and so on	Learning Center
Training Timetable developed	By FAMTs, GJTs, local job trainers, and leaders	To have a clear plan for the training of local employees	Learning Center and Lean Office
Local job trainers train workforce	Classroom training and shop floor training	To have everyone trained according to training plan	Learning Center and Lean Office
Local job trainers conduct follow-up until workers follow process consistently	Follow-up by trainers on employees until employees master the skill	To ensure the employees have learned the skill	Learning Center and Lean Office
Learning Center evaluates trainers' skills	Observation of local job trainers training employees	To sustain a high level of trainer skills	Learning Center

Establishment of Follow-Up and Process Confirmation Functions

The establishment of follow-up and process confirmation functions was essential. These were developed to ensure that employees had learned and mastered the skills. The trainer conducted a follow-up session with the employee concerning the skill in question, after which the leader of the area took over and went through the process confirmation with the employee to understand whether more training was needed. If the leader found the skill set inadequate, the leader had the opportunity to find another job the worker could master.

What	*How*	*Why/Target*	*Support*
Leaders are trained in process confirmation (on standard work)	Introduction course on Lean	To ensure that leaders can conduct process confirmation in the correct way and understand why	Lean Office
Leaders conduct process confirmation on workforce	Process confirmation on employees conducted by leaders	To ensure the employees have learned the skill and are not put on a job they have not mastered	Lean Office
Local job trainers follow up until workers follow processes consistently	Follow-up conducted by trainers on employees until employees master the skills	To ensure the employees have learned the skills	Learning Center and Lean Office

Establishment of Continuous Improvement Structure

To ensure involvement of everyone in the development of the best standards, when the first employee was trained or when follow-up and process confirmation on a job was conducted, the trainer or leader would ask if there were any improvement suggestions regarding the job. The success of Lean can be measured by the speed with which we move from improvement suggestion to test of suggestion to evaluation of suggestion to input back to proposer and, finally, to implementation of suggestion if approved.

What	*How*	*Why/Target*	*Support*
Establish local trainer information boards and weekly routines	Introduction course on Lean	To ensure that the motivation to give improvement suggestions is high and the results are visual to all	Lean Office

Continued

What	*How*	*Why/Target*	*Support*
Establish global weekly continuous improvement LiveMeetings	Weekly LiveMeetings	To ensure that the motivation to give improvement suggestions is high and the results are visual to all	Lean Office
Ensure that proposer always receives feedback (quickly)	Feedback from employee to trainer to employee	To ensure speed, which is our success	Lean Office
Hold routine meetings between leaders, workforce, and trainers	Weekly job training huddles; FAMT has monthly meetings with the leaders (1 on 1)	To ensure that there is a clear link between production and training	Lean Office

THE GLOBAL PILOT PROJECT

Chapter 4

Testing the Global LEGO Training Organization and TWI JI

Introduction

As discussed in Chapter 2, the consensus was to run a pilot project on the following task: Mold change for VC Engel 60-ton molding machine, with a three-plate mold, working with ABS-type plastic. We constructed the project model in such a way that we would hold workshops every 4 to 8 weeks, and the pilot project team would meet at all of the plants equally so that no one plant or group would be unfairly represented (see Figure 4.1). At the same time, we would conduct a thorough evaluation between each workshop and allot time for preparation and work assignments that would be done at home in their local plants.

As explained, we came to the conclusion that splitting the workshop into two working groups, the Operations Track and the Learning Center Track, would be the most efficient way to move forward, giving team members in the two tracks the opportunity to work independently on similar topics yet the ability to participate together in group events. The division of the two groups allowed us, then, to focus separately on "operations" and "training," while providing the necessary overlap for everyone to develop and practice the Training Within Industry Job Instruction (TWI JI) skills together.

While both teams were meeting to learn the TWI methodology, the Learning Center consultants would work further on, and make

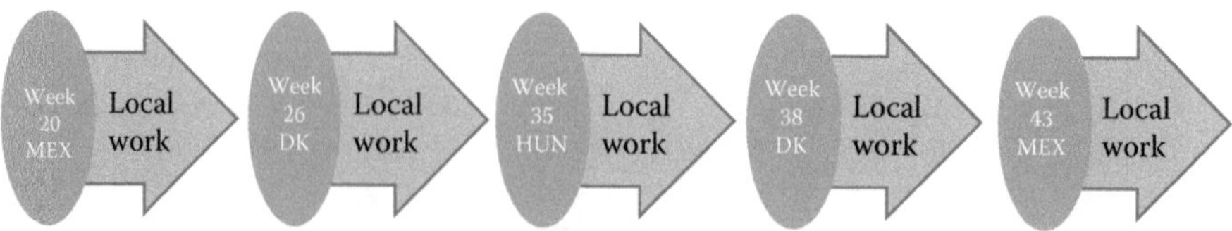

Figure 4.1 The pilot project model: workshops and "homework."

adjustment to, the tools and processes for development of the ultimate global rollout of the training organization. This meant that while the operations team worked on the details of breaking down and training the mold changeover process, the Learning Center team would be working at a higher level, creating the structure for recruiting and developing trainers who would promulgate this training process throughout the organization. Having these Learning Center consultants working together with the experienced practitioners and experts meant that the test team would be involved in the process of developing these tools; this was an extremely important part of the pilot project model. Otherwise, we would run the risk of it being a tool developed in the back office that was not grounded in reality and ultimately not used.

Another important aspect of this dual-track structure was to have the operations team supported by more than one consultant from different plants to gain breadth in the testing and preparation of a strong rollout. Using multiple consultants, and having more than one set of eyes on any one issue, created a stronger rollout phase and a stronger structure at the finish line. Moreover, because the consultants were coming from different countries and cultures, we could ensure that the final product would be viable for all of the plants in their native environments.

By spreading this work over several departments from the various plants, we were able to have more employees taking *ownership* of the project, which amounted to greater unity of purpose, defying cultural boundaries and borders. This created a strong foundation for future trainers to build on and create standardized work. We believe that one of the biggest stepping-stones in the change process lies in ownership. For employees to embrace this change in culture, they must not feel that it has been forced on or rushed at them, but that they have taken part in shaping the new culture and that their suggestions and objections have been heard.

Throughout this project, we strove to keep a strong focus on change management principles to allow the organization to undergo a change process that was slow but steady. We even studied together the dynamics of the change process so that we would be aware of and prepared for dealing with the resistance that naturally occurs when people go through a major change, whether at work or in their personal lives. We realized that one of the most important, perhaps the most critical, element in ensuring the success of the change process was management support not only to ensure the necessary resources for the project but also to encourage and promote the change that needed to take place. We especially did not want to repeat our previous experience, in which training projects failed because of a lack of management support.

Reactions to the Change

One of the first challenges we faced when starting the pilot project was motivating and creating confidence in the members we had selected to be part of the team. This advancement and change from being a local member of the workforce to now participating as a key player in a global project, whose sole purpose was to change the organization's learning and training culture, generated a lot of pressure on the participants. They needed to embrace the new pilot methods as well as see themselves in a position where they could work alongside the Learning Center consultants in the implementation phase of an entirely new set of tools. In short, they were asked to change their own methods of work while promoting the new methods they had just learned.

When we were all gathered together as a team during the workshops, we were able to talk and discuss openly about this change process because we were all in the same boat and understood the challenges and the value of the project. But, at home, during the breaks between our meetings, it would be a different story. Some of the team members worried about the reaction they would receive from their colleagues on the work they were doing and their role in it. Responses from coworkers and even management were critical in this phase, when locally the plants were only slowly starting to understand and accept the new paradigm. This added to the growing insecurity the team members felt about the validity of the work they were performing within the project team.

We did our best to support and help the positive efforts of the team, but we were afraid that after completion of the project, we would not be available to help in the same manner. This is why building a solid groundwork with this particular group was essential. Also, within the local management group, it was critical to build a solid foundation so that the group would give continued strong support to the trainers after the project was completed. There were many difficulties in doing this, especially with a project such as this, for which results would not be seen right away but would take time and patience to build.

At times like this, things can become problematic because the urge to obtain quick, substantial results is overwhelming, as opposed to sticking with the program and creating a sustainable process that involves all employees. That, after all, demands so much more time and effort. But, when management makes some room, as the LEGO management allowed during this project, and leaves it to the trainers to develop and share their knowledge and skills, they are signaling a high degree of trust, recognition, and appreciation of the trainers' knowledge and capabilities. In addition, they were also empowering them to raise the competency levels between their coworkers by training them and implementing their suggestions. This process will stabilize the whole work environment and will place a higher focus on acknowledgment and development, which will give a positive effect on production. This is the way we saw it in the Global Learning Center: "To create a learning organization takes time and hard work; it's not a sprint, it's a lifelong marathon."

Learning How to Make Job Breakdowns

The vital part of the pilot project effort was the development of Job Breakdown Sheets for training that would be created by the global team to understand the pilot jobs and prepare for training them. Since all members of the pilot project team were to become trainers they needed to learn to use this tool, the Job Breakdown Sheet, which came with the TWI JI method. It is an analysis of the job to be trained that "breaks down" the job into the elements of Important Steps, the "what to do" of the job; then Key Points for each of those steps, the "how you do it"; and finally the reasons for each of those Key Points, the "why we do it that way." This breakdown is then used during the training to be sure the trainer covers the critical aspects of the job, in order, without forgetting anything. Although at first glance this analysis may seem straightforward, in reality it is a sophisticated tool in which the trainer must carefully select which details to include and which to leave out so that the learner will obtain just enough information to be able to quickly remember to do the job correctly, safely, and conscientiously without becoming confused or overwhelmed with too much information. We began making these breakdowns immediately after the trainers passed the TWI JI 10-hour class. The first drafts of these breakdowns were then developed during the second workshop, and then they were developed and refined over the next several workshops and homework periods.

It became critical for us to develop the Job Breakdown Sheets in the correct manner from the start of the process. This was done to ensure that the trainers quickly developed the skill sets needed to create a good Job Breakdown. It also ensured that the trainers had a feeling of true ownership of the Job Breakdown Sheets and developed and fine-tuned a common terminology and language around the jobs we were to handle in the pilot.

Before starting to facilitate the process of making a breakdown, we set up beforehand a standardized format for the environment in which this effort would be done. This meant setting up the working area where breakdowns would be made with a whiteboard, which was divided into columns for Important Steps, Key Points, and Reasons, just like a Job Breakdown Sheet (see Figure 4.2 for an example of a Job Breakdown Sheet). We also ensured that the people in the specific work area where we would be observing

LEGO **Disconnect Mold Temperature Controller**
Job Breakdown Sheet

Created by: István N., Laszló Zs, Bent J., Tommy H., Gustavo O., Alberto R.	Document Owner: Klaus A. N., Marien P. V., Flemming L. K.	Date of Implementation: 06.04.2012. Date of Change: Estimated Training Time:
Parts:		
Tools & Materials:		

Nr.	Important Steps	Key Points	Reasons for Key Points
	A logical segment of the operation when something happens to advance the work.	Anything in a step that might— 1. Make or break the job 2. Injure the worker 3. Make the work easier to do, i.e., "knack", "trick", special timing, bit of special information	Reasons for the key points
0	Common key point	1. Follow general safety rules	
1	Close flow meter no. 4	1. Closing water battery 2. Closing both valves 3. Closing them finger-tight	1. Close down flow meter 2. Releasing the water 3. Not to damage the threads
2	Turn off and release pressure (LOM, NYI)	1. Changing to vacuum mode	1. Risk of water on the floor
3	Turn off the power (BLL)		
4	Disconnect all hoses	1. Pulling the security cylinder 2. Placing the hoses up off the floor	1. It won't come off 2. So nobody falls over the hoses

Figure 4.2 Example of Job Breakdown Sheet.

the job and breaking it down knew exactly what was going to take place there and why. Furthermore, we made sure that all the needed parts, tools, materials, and machines were available for the breakdown sessions. All these small actions made the team feel their work was special and created a professional atmosphere.

We were taught in the TWI JI 10-hour class that the development of the Job Breakdown Sheet must be done on location where the job takes place. This is a necessity in order to observe the most realistic environment for the process; in this way, you obtain a hands-on approach that facilitates the process of seeing the Key Points. This cannot be replicated in a classroom away from the job site.

This was one of the mistakes we made between the first and the second workshop: We tried to make the Job Breakdowns sitting behind a desk using only a computer. And, this is where the process came to a halt. Not being on location created a mental gap, and we were not able to describe a complex procedure from memory. This made the first Job Breakdowns a failure as they were missing Important Steps and Key Points, and they were too "wordy." In retrospect, this last point resulted from the fact that we were not standing in front of the machine and hence could not see how simple the command should be using the *active voice*. Instead, we tried to explain it in detail, the way we saw it in our heads, as if we were writing a book.

This just shows how easy it is to "get lost" in old habits. The old procedure for the development of training material had always been seen as a "behind-the-desk job," which is why the trainers, at first, regarded the new method as a waste of time. Why perform all that work for a simple task? Just sit down and write it out. But, it turns out that what we think to be a simple job is really much more involved once we take a good look at it and analyze it with a strong method like the Job Breakdown. All these small things reiterate that learning TWI is not just classroom training, an online course, a webinar, or something you read in a book—you have to practice it at the work site with a mentor to learn it.

Once we had the team gathered around the machine where we were going to make a breakdown, we simply followed the script for the Job Breakdown process that we learned in the JI 10-hour course. This meant we started out having one of the team members performing the job. Once the volunteer had performed it two to three times, so all could see and understand the job, and we had everyone agree that this was the way to perform the job, we would ask the team whether they were ready to start defining the Important Steps.

Facilitating the Development of Important Steps

The facilitator would begin by asking the volunteer to perform the job one more time without saying or explaining the procedure. It was now a group effort to identify the first Important Step as something that "advances the job." This was simply done by saying "stop" when we thought the volunteer reached a point during the procedure at which the job had physically or mentally moved forward, becoming closer to completion. When one of the trainers indicated that such a point was reached, the facilitator then asked, "What has the volunteer done?" It was now the team members' job to articulate exactly what the volunteer had done. "Has the job advanced?" would be the next question for the facilitator to ask. If the team members said, "Yes," the facilitator would follow up by asking, "Can this be an Important Step?" If the team members again responded "Yes," that implied it was an Important Step.

The facilitator then wrote the Important Step on the whiteboard, as the members had stated it, and then turned to the group and asked, "Do you agree with this being an Important Step?" While this methodology seems cumbersome and perhaps redundant in places, by going through the discipline of having all the participants perform the process repeatedly for determining each of the Important Steps, the facilitator was drilling them in the process of making the breakdown. In other words, we were creating a habit for them to do it correctly each time they did it.

Oftentimes the participants entered into vigorous discussions about the wording of the Important Step or whether it was the right place to stop, but this was healthy because it created a strong ownership of the process by the whole group. It also deepened and strengthened the learning process, which would make it easier for them to continue performing the breakdown process effectively in the future. The wording of the Important Step makes a difference to the quality of the training, so trainers have to have a clear understanding of who they will be communicating with and use terminology familiar to this group as opposed to technical words or trade jargon. Engineers are not the ones who will be operating the machines, so we had to remember the importance of seeing eye to eye, or "ear to ear," with the worker.

When the participants had agreed on the first Important Step, the facilitator would go through the same questions again to define the next Important Step and continue this loop of questions and answers until the complete job procedure had been covered. During the process of determining the Important Steps, it was a good rule of thumb to make sure that everyone agreed and that all had a chance to be heard. There was nothing worse than having to step backward in the process just because someone was not paying attention and then changed his or her mind.

Facilitating the Development of Key Points

Once the Important Steps were all determined, we then moved on to find the Key Points for each of these steps. Before we looked for any one Key Point, though, it was imperative to have a clear understanding of the conditions that make a Key Point so that we knew what it was we were looking for. Therefore, the facilitator continually asked the group, "What are the three conditions to be a Key Point?" The trainers recite the conditions: (1) Does it "make or break" the job? (2) Does it "injure the worker"? or (3) Does it make the work "easier to do"? This might seem a pointless exercise, but because we were learning how to find Key Points, this repetition was an essential part of building the skill of making breakdowns. It created focus and locked in the attention of the trainers to critical aspects of the job that are taken for granted or done unconsciously. We never encountered anyone who was able to perfect the skill after only one repetition.

The facilitator would then ask the volunteer to perform the job, again without saying or explaining anything, following the Important Steps one by one. It was still the trainers' job, as they watched each Important Step performed and looked for anything that might meet the three conditions, to say "stop" when they came across what they thought might be a Key Point. When a participant found something and pointed it out, the facilitator would ask: "So, why did you do it that way?" or "What would happen if you did not do it that way?" It was then the responsibility of the volunteer doing the demonstration to answer this question by explaining the exact reason for the Key Point. The facilitator followed up by asking: "Then, which of the three conditions does it meet?" and the volunteer identified at least one of the conditions met by that Key Point. That would confirm it was, indeed, a Key Point. The facilitator then asked the volunteer: "What should I call this Key Point?" and wrote it on the board together with the mentioned reason. The facilitator would then turn to the rest of the group and see if they agreed with the Key Point and the Reason for the Key Point. After gaining consensus from the group, the facilitator continued in the same manner to find the rest of the Key Points along with their reasons for that Important Step until no more were found.

The process then was repeated for the next Important Step and then all of the Important Steps until the end of the job.

When you read the procedure delineated, bear in mind that this is a training situation in which we were learning how to make breakdowns, so repetition and understanding were the keys to success. So, the more specifically a job function was narrowed down, the better. And, the more times it was repeated by the trainers learning to use the method, the easier it would be for them to train others on the job in the same exact manner.

When all Important Steps, Key Points, and Reasons for Key Points were defined, described, and added to the whiteboard, the facilitator took a picture of the board and sent it to the technical writer assigned to the workshop. The technical writer created a digital version of the Job Breakdown Sheet and printed copies for all the team members. This normally took only 20 minutes or less, during which time a much-welcomed lunch might take place.

After the Job Breakdown Sheet had been distributed to all of the workshop participants, the facilitator approached the group again, asking whether there was anyone willing to teach the job to another person. This volunteer would teach the job using the TWI four-step training method along with the newly developed first draft of the Job Breakdown Sheet. When teaching the job, it was not unusual for the group to discover new Key Points, to combine or divide Important Steps, or to shorten the phrases used on the breakdown sheet. In other words, the breakdown sheet was always a work in process that could and should be fine-tuned and adjusted after seeing the results of how it worked while actually training an operator. After the whole job training process had been carried out three to five times, the adjustments we discovered would be implemented with the full consensus of the group.

It was only at this point that we felt we had a good starting point for beginning the "real" job training—that is, training operators to perform the job. But, having done all this work, the team was more skilled in developing new Job Breakdowns Sheets on their own, and we learned that follow-up coaching on the TWI 10-hour JI class was, in fact, a necessity.

Finding Key Points: A Great Learning Experience

We learned from making the breakdown sheets together as a group that identifying the Important Steps and agreeing on them was a good icebreaker exercise. It was a different story, though, when going deeper into the job to look for and find the Key Points. Here, the group sometimes took a long time defining and agreeing on even just one Key Point. We normally say that the devil is in the details, and here the details *are* the Key Points.

A great example of not only how challenging it could be to develop a single Key Point but also how valuable the time spent was to both the learning and to the operation itself occurred during the Job Breakdown for the *Mounting Mold* task. When mounting the mold to the machine, one of the Important Steps was *Hang Mold* to the fixed side of the machine. This meant that the mold, weighing approximately 800 pounds (350 kilos), would be hanging from the fixed side attached by four bolts, two bolts on each side, one at the top and one at the bottom of the mold.

When performing the task, you would start on one side of the machine, the front side, and attach the first two bolts, then go all the way around to the other side of the machine and attach the two remaining bolts on the back side. Then, you would return to the front side of the machine to adjust and fine-tune the mold, only to have to walk around again to the back side of the machine to adjust and fine-tune the mold on that side. Finally, you would have to walk yet again around the machine to return to the front of the machine to continue the job. A lot of walking around the machine back and forth took place.

During this session, one of the global job trainers (GJTs) from Denmark argued that we could save a lot of walking, and time, by mounting the two bolts on the front side and then completing the adjusting and fine-tuning of the equipment on this side before going to the other side to attach the last two bolts and to complete the job there. This spawned a stir within the Mexican group as one of their global trainers replied: "No, we cannot mount only two bolts and then release the crane hook to adjust and fine-tune the mold. The mold is too heavy, and the bolts will break; the bolts we use in Mexico are of poor quality!" We were dismayed by his opinion because we wanted to establish the same Key Points in all of the training material, no matter where we did the job, to create greater unity and less variation between the different sites. And, of course, we were puzzled that the Mexican bolts apparently were of inferior quality.

We therefore started out by investigating whether the three factories did indeed use the same type of bolts. They did. The next step was to examine how much weight the bolts we were using could hold, and it turned out they could hold up to 10 tons each, much more than the weight of the mold. This was a true mystery. We had an 800-pound mold mounted with bolts intended to hold up to 10 tons, yet the bolts were still breaking.

Then, we had a breakthrough. One of the Danish trainers remembered that quite a number of years ago they had encountered a similar problem with the bolts holding the molds. What happened then was that at the plant in Denmark some of the mold setters had been overtightening the bolts when mounting the molds, which then stripped the threads on the machine, which meant that the mold was basically just hanging loosely from the machine. This is a hazardous condition should the mold fall, threatening the safety of the operator as well as the value of the mold. It was all a matter of the force put into tightening the bolts and how much pressure should be applied to this specific action.

This new knowledge, unfortunately, created some consternation with the Danish trainers as they had been responsible for the training of the Mexican setters during the startup phase and had left out this vital piece of information. But, there was no point in criticizing or pointing fingers at whose fault it was, and the energy quickly shifted to finding a sustainable solution to the problem at hand. The real problem here was the training. How do you describe the right amount of pressure or the correct feeling when turning the bolts? This is just the kind of skill covered in the Key Points found using the TWI methodology. We used the Key Point as a reminder for the trainer, when teaching the job, to show and let the learner feel how to tighten the bolt just right. Best of all, we could all then agree that the way *everyone* would do the job would be to hang the mold with two bolts and complete the work on each side of the machine, eliminating the need to walk back and forth multiple times.

This is a great example of the many lessons learned in finding just one Key Point, and it took us about 1.5 hours to develop.

First Workshop: Hungary

The first workshop was planned to be held at the LEGO plant in Nyíregyháza, Hungary. One of the main goals for the workshops was to eliminate, finally, the idea that, "Denmark is always right." Another issue that played a part in choosing Hungary as the first meeting place was that much focus and attention were already being given to the newly built plant in Mexico, with numerous meetings held there and people from around the world, especially Denmark, going to Monterrey to help them get started. So, it was decided to have the first workshop in Hungary so the members of the team there would not feel they were being left out.

Then, the unthinkable happened. The Icelandic volcano Eyjafjallajökull erupted just days before the first workshop was supposed to kick off.

The ash cloud created by the volcano stopped most air traffic over large parts of mainland Europe and made it impossible for any of the team to get to Nyíregyháza in time for the workshop. The team consisted of 15 individuals from five different countries, as well as the TWI trainer, Patrick Graupp, coming from the United States, and there was no way to reschedule or postpone it even a week or two.

So, the whole first workshop had to be cancelled at that point. We then pushed the first workshop back to what would have been the second workshop, which was to take place at the LEGO plant in Monterrey, Mexico. The whole team was already planning for this trip to Mexico, so the pilot project started one month later than scheduled.

Imagine our frustration: All the planning and energy for the initial kickoff had gone to waste because of an ash cloud. It just showed that you cannot control everything in life.

The Workshops

The full pilot project team had its first kickoff workshop in Mexico in week 20 of 2011. The full working team (not including the sponsor, the steering team, or other stakeholders) consisted of the Learning Center manager; the project manager; one Learning Center consultant from each plant (Denmark, Mexico, Hungary, and the Czech Republic); one functional area master trainer from each of the three molding plants (Denmark, Mexico, and Hungary); and two global job trainers (GJTs) each from the three molding plants for a total of 15 people. In the next chapter, we go into the detailed contents of each of the workshops that were held during the course of the pilot project period. First, let us look at some of the common aspects of all the workshops.

Each weekly workshop started out the same way on Monday morning with a short preparation session so the team members could relax and focus on the content of the week's work. The first 30 minutes were used for the team to reacquaint themselves with each other and change into their mental "work clothes." We achieved this by going over the plans for the week, which were created specifically for each of the two tracks: the Operations Track and the Learning Center Track which were managed separately by Vellema and Jakobsen. Each part of the schedule was described in terms of "input–process–output," so we all knew what was expected of us.

This was followed by a 15-minute welcome and presentation by a local member of the steering team. These presentations focused on the ideas and expectations for the pilot project from the local point of view. In other words, they gave an opportunity for the local plant to address the team, right at the start of the week, on their hopes and aspirations for what the team was trying to accomplish. It was inspiring to see the various steering team members greet the pilot team with open arms and express why they thought the project needed to be successful and why it was important to their specific plant. We always made sure to brief the steering team member before his or her introduction to make sure that they covered the same topics, just with a local twist, and to see if other stakeholders wanted to speak to the team. In fact, their welcome speech was often followed by a short talk on a chosen subject that the local human resources director or Lean director felt would provide value to the team.

In this way, we always secured a good start to each of the pilot workshops and kept the local stakeholders aware of the presence of the team during the workshop week. This awareness ensured that these stakeholders would stop by several times during the week to make sure that everything was going according to plan, but also, more importantly, ensured their show of support and enthusiasm for the pilot project in general. This not only encouraged the pilot team but also indicated just how serious the pilot project was being taken in the local facility.

The workshop weeks were arranged in such a manner that the local pilot team would act as hosts for the workshop. This was a great help when it came to preparations because all the practical issues could be arranged locally, and a social event, which took place Wednesday after work, could be planned. By letting the local teams handle these responsibilities, we showed them that we respected their knowledge of the local community, their country, and their plant. And, by doing so, we made sure that the local team took ownership of the workshop, and that they took the necessary time to prepare the local stakeholders. Each plant's team made a big effort when it was their turn to host, and we could all appreciate the pride and joy by which they showcased not only their home plant but also their country's cultural treasures. This made these workshops something looked forward to by every member involved.

Cultural Introduction

During the initial phase of the workshops, we wanted to establish a good collaboration across cultures between all of the plant teams. This is why we decided, as a small beginning to cultural training, to show the teams various videos about culture and the miscommunication that might occur when non-English personnel communicate in English. This inspired a constructive dialogue between the different site members, which showed us that everyone wanted more information on the subject of culture and found it not only necessary but also interesting. When we discovered that the teams were eager to learn, we decided to try to find funding to invite an external consultant, with expertise in cultural differences, for the next workshop.

Social Event

During each of the workshops, the local host team was put in charge of arranging some "entertainment" for a Wednesday night social event. The bottom line for these events was to have time off from the work mind-set and give the team a chance to interact on a more personal level to know each other better. We felt this was essential if we were to establish a positive working environment for the entire group. What is more, by having the host team come up with some way to introduce the participants to their country's unique characteristics, we could advance the culture study to local sights, sounds, and tastes in a way that built respect and admiration on the part of the guests and pride and self-satisfaction on the part of our hosts.

Furthermore, we knew that most of the team members were regular, hardworking, hourly paid employees who might never have dreamt of traveling across the globe to visit these countries. For them, it was a big deal just to participate in a global project and to fly overseas. In the end, this is what bound the team together, knowing that at some point they also would be traveling to an unfamiliar country and would need to depend on their international colleagues to take care of them in a strange and new environment. This gave everyone some common ground to share and talk about in addition to some time outside their own plant. In this space, they would confide in each other and strengthen their relationships.

That was the intention. And, these events did, in fact, strengthen team spirit. However, it was quickly seen that when the members were given this break and had time to enjoy the new surroundings, they would without any prompting be able to solve difficult tasks that were presented to them earlier during the day or week. We see now that this "time off" in a relaxed and informal environment actually allowed them to speak more openly about the challenges they were facing at that point, and this may have been the factor that spawned fresh thinking and out-of-the-box solutions.

"Homework" Periods

The project model was built so that there would be periods of time between the workshops held at the three international locations for team members to complete "homework" assignments before the next workshop. The idea was to have approximately 4 weeks dividing each workshop week. This meant that the pilot project would take half a year to complete. The reason for these four weeks of "home time" between each workshop was that we knew that the members of the group would need time to reflect on all the material they had learned. By giving them this time off and specific assignments to complete, they would be able to transfer the learned material to real situations in their local environment, thereby enhancing the learning process. This is basically an "expose-extract" procedure, meaning that they would be *exposed* to new material and then have the time and space to *extract* this new material and transfer it to their daily routines. This is why we only allocated 50% of the GJTs time for the project, so that on return to their site they would be able to test the new material while keeping up to date with the work and in touch with their coworkers.

The assignment for these home-time periods was for the members to develop a draft for an assigned Job Breakdown that was decided on by the GJTs during the workshop. The different sites were assigned different jobs to break down depending on their areas of expertise, particular aspects of the work that were being emphasized in their plants, their own personal experiences, or other factors. They were basically assigned homework after each workshop. After the draft was made, it was sent to the other sites, where it would be tested and suggestions for improvement developed.

At weekly videoconferences facilitated by the Global Learning Center during these off weeks, the GJTs would present the Job Breakdown their team had developed and hear the ideas for improvement the other sites had made. Some of the TWI Job Breakdowns could take up to four videoconferences before they were complete, and we could, in general, talk about one to four different breakdowns at a single videoconference. In the first homework assignment period, we found it remarkably difficult to develop a good Job Breakdown. They became too long, and we kept confusing Important Steps with Key Points and the order they should be developed. This called for a swift solution. We agreed to bring in the senior master trainer of the TWI Institute (Patrick Graupp, coauthor of this book) to conduct a 3-day TWI JI follow-up session, the one Robert Wrona had already advised us to take at the very beginning. So, his advice turned out to be correct.

It was one thing to have trouble developing even the simplest Job Breakdown, but what was a completely different and unexpected issue was to have our team members argue over each other's viewpoints during these first videoconferences. This hindered our effectiveness a great deal. A growing frustration started to evolve: Would we ever succeed at global knowledge sharing? It turned out to be a lack of cultural understanding and lack of new relationship development with all of the members. For the most part, it was a basic issue of trust, and as the workshops progressed and the group congealed, the videoconferences progressed more smoothly and were more effective.

The fact was that the more the team met in real life and socialized, the closer the bond between them grew and the better the process became. This was clearly proven by the development of a common terminology around the Important Steps, the Key Points, and the Reasons for the Key Points. In this way, our work became more tangible, and we changed the focus away from who was right or who was wrong toward the facts and the rules behind making the TWI Job Breakdowns. We recommended that the GJTs meet, even after the pilot project was completed, a minimum of twice a year to maintain this good relationship and keep up the work effectiveness.

Chapter 5

The Workshops

Introduction

So far, we have given a general background to why we set up the global workshops and how we went about doing them, but when it came down to actually bringing the people together and experiencing the human dynamic of the interaction, we learned a lot about how to make this whole thing work. In this chapter, we walk you through the details of what actually happened when we tried to turn our plans into action, the problems we encountered along the way, and how we solved them. Even the best-laid plans and strategies must ultimately come face to face with the reality of what it takes to actualize them and produce results. So, this is the story of the activities we organized and carried out in the pilot project.

Workshop 1: Mexico

Finally, we all gathered at the factory in Monterrey, Mexico. After the standard introduction on Monday morning, we had a team kickoff session where we went through the scope of the project, the project plan, the job training concept, and the Learning Center mission and vision. Furthermore, it was important for us to emphasize to the group how we had handpicked them especially for the pilot, and that the expectations for the outcome of the pilot would be the foundation for the LEGO Group's new learning organization. We did this to make sure that when Friday came around, they would have gained insight into the ideas and reasons behind the project and would thereby take ownership of it.

The main objective of the first workshop was to get to know one another both in a professional manner and on a more personal level. The latter part might seem irrelevant as we were dealing with training and learning within industry, but we could not neglect drawing the team members into a cohesive group as the foundation on which to build our project as well as the overall organization. Understanding the motives for people's actions is of essential value when it comes to any type of cross-cultural training. Because these were the people who would build and cocreate the new Training Within Industry (TWI) Job Breakdowns that we planned to use for training across the entire international organization, this meant that they would be working closely together in intensive workshops.

Knowing what makes a good trainer was extremely relevant at this early stage to create a harmonious work environment. That said, there will always be big personalities colliding with each other when different cultures meet, and we knew not to take this as an obstacle but as an advantage. The more they tended to argue, the better it would be. That just meant that dedicated people had been selected who truly wanted to show their worth and make a strong contribution. With the right people on board, they most likely would arrive at a compromise after investigating all possible scenarios. We would have been more worried had we *not* selected people who were as strong willed as the people who came to Monterrey that week and met each other for the first time.

The first point of business at the workshop in Monterrey was to begin making the Standardized Work Chart (SWC) and the TWI Job Breakdowns for the designated pilot project jobs. This, however, quickly came to a standstill when we discovered the lack of a "common language." In other words, we needed a common understanding of what things like, "Grab hold of this," meant as opposed to saying, for example, "Grab this." Believe it or not, with the different native languages and cultural viewpoints, simple differences like these created confusion and misunderstanding when everyone was communicating together in English. This meant that having a common language and terminology was essential to building a global training organization. This realization brought out the following points during the first workshop:

- Understanding each other's point of view had to be made a priority.
- The available standard operating procedures (SOPs) were below par, and their translations were poor.

When we went over the team member roles, assigning each of the participants their duties for the project, we confirmed for ourselves that we had indeed done the right thing by not taking the easy path during the selection process earlier in the planning stages. As we pointed out in Section I, we spent many months developing the roles and finding the right people to fill those in the pilot. Because we had clearly defined the expectations up front, we had a good image of who might fill each role and therefore were able to recruit the right people and received a heightened commitment on each of their parts. This was something we needed to establish immediately in this initial phase. All of the preparation now looked like it would pay off when we had a chance to meet these people face to face and see what each of them brought to the project. The roles were distributed across borders, meaning we would have people in each country performing the same tasks in parallel at the different sites. They were equally filled across the plants, and all of the team members had a clear understanding of what their roles entailed and why. This gave us a sturdy platform for dialogue and openness toward one another.

Operations Track (Functional Area Master Trainers and Global Job Trainers)

As explained in Chapter 2, we divided our pilot project team into two tracks, or two distinct groups, one focused on operations and the second on the learning aspects of the project. In the Operations Track, the first days went by rather quietly; this is normal for any new group dynamic. Here, the team was just anticipating each other's moves, slowly getting to know one another, and finally learning what to expect from the workshop. We had to keep focus on the fact that the three sites would need to work together to create the best results. In the past, production knowledge and know-how had always been "exported" from the Danish site, and it was seen as inappropriate to question this knowledge. Our current approach was aimed at the international sites taking part in the action and giving their input to the work procedures, when they felt ready to do so, as well as for the Danish site to start listening and learning from a fresher perspective.

We quickly understood the necessity of working with something specific, concrete, and practical to showcase the benefits that a systematic approach gave to the responsibility of sharing know-how and skills. Learning from each other and sharing tacit know-how was at the core of the workshops, so basically our assignment was to turn something intangible into something tangible. We therefore decided to have each plant's team perform the *changeover process* of the pilot case to ensure that the current method was documented and could be used for comparison later, not only at a local but also at a global level. Comparing results of the before and after processes was a good idea to showcase the results of the pilot project to upper management, and it would help motivate the team when it came to future implementation.

This was easier said than done because it was a puzzle at first how to actually re-create something from one country to another without moving the whole operation. The solution was simple. We decided to videotape all of the operations down to the smallest detail. The idea was first to have the Danish trainers show and tell how they would make the mold change, then the Hungarians, and then the Mexicans, at the same time recording the proceedings while the other participants watched the job proceed and took notes for later use.

This turned out to be the ideal solution. We quickly saw the variation between the different plants and their basic operations. There were also differences in the attitudes toward safety, quality, and the time needed to perform the changeover. Two of the more critical examples were as follows:

Temperature Measuring. One of the plants had chosen to completely disregard a key operation in the molding process: measuring the temperature of the material. This is a critical item as the temperature of the material determines the quality of the end product. The first reason they gave for skipping this important step was that it consumed a lot of time. It does take time to measure the temperature, but in the end, it will save time because you will have to adjust the temperature during the production cycle because of too little or too much heat.

The second reason given was even more critical to the workshop: It was lack of knowledge. The workers did not know about the importance of an exact mass temperature and did not know the "why" regarding the operation. So, this proved that there had to be a communication gap somewhere within the learning and training.

Safety. The second example dealt with one of the plants having a high focus on safety during the changeover procedure. This included, among other things, sweeping the floor to reduce the rate of accidents that were often seen in relation to slipping and falling on waste material from the molding process. This started an intense discussion about safety versus time.

> It was a general understanding, or viewpoint, that efficiency equals speed. But, this is not always the case. We saw efficiency as something that relates more to correct, safe, and conscious procedures rather than just speeding the production pace. This was a change in attitude toward the whole concept of being "effective." We had to start out with the whys of the operation because knowing the "why" of an operation gives insight into the importance of each procedure performed. The reason for sweeping the floor was not only to keep the area tidy but also to prevent reoccurring accidents that were proven to be caused by material left on the floor.

The lesson learned from these two examples was that if workers have not received clear instructions or training on how to do the job correctly, they will invent their own procedures to meet their objectives. The fact that we actually had the opportunity to have discussions on topics like these was a great way to start our project, not only for the outcome in general but also so the participants would have something concrete to bring home with them. What is more, by actually videotaping and viewing the procedures, we could deal with facts (what we saw) and not simply argue over who was right or wrong. This paved the way for open and productive discussions and gave us a good model to use as the project moved forward.

Standardized Work Chart

We had arranged an introduction to the SWC, a Lean tool we wanted to use as the basis for our work. This was a higher-level document to the Job Breakdown that would outline the work in its entirety using detailed job descriptions (work elements), symbols to highlight special features (especially safety issues), layout drawings, and other details such as timing and distance (see Figure 5.1). We held a videoconference with a Lean consultant from the Lean Office in Denmark. This rather simple introduction worked well as a catalyst for a dialogue on how to link TWI's Job Breakdown with the SWC and get us started developing an SWC for the pilot case.

Having videotaped the three different changeovers of the mold and having received the SWC introduction, the Operations Track team started developing the SWC for the pilot case. We began by having each of the three sites develop a draft based on their own work. Then, in collaboration across the sites, we started making a preliminary draft of a new SWC that was a consolidation of all the sites' work, a "one best way" we could

LEGO

Cleaning material hoses and connector

Standard Work Chart / Standard Arbejdsskema

Plant Billund	Created by:	Approved by:	Walking	Return to start	Standard in process stock
Department Kommarken	Rene F. Kristensen	Tom Vinther Kristensen	→	- - - →	●
Module / line: Change over					

Flow	WES	Work Element Name	Symbol	Manual	Auto	Walk
1	04.03.784961.WE	Cleaning of hose by grinder		55	-	Sec.
2	04.03.144109.WE	Cleaning of grinder hose **by maguire**		48	-	2
3					-	
4					-	
5					-	
6					-	
7					-	
8					-	
9					-	
10					-	
11					-	
12					-	
13					-	
14					-	
15					-	
16					-	
17					-	
18					-	
19					-	
20					-	
21					-	
22					-	
23					-	
24					-	
25					-	
Takt time:		Sec.	Sum	103	-	2
Cycle:		Sec.	Total: (Manual + Walk)	105		Sec.

1

2

Document name: 04.03.010068.SW Cleaning material hoses and connector.EN
Document owner: Tom V. Kristensen

Figure 5.1 Example of Standardized Work Chart (SWC).

all do the process. As a starting point, we determined that each of the work elements defined in the SWC were to become a headline for individual Job Breakdowns for training. The team regarded the SWC as the "table of contents" for the Job Breakdowns. Later, we realized that creating this tight a structure was not always necessary, but it was a good way for us to start and to make the connection between the two tools. By the end of the first workshop, we had the first version of the SWC completed for the pilot case.

TWI Job Instruction 10-Hour Class

Following our plan, we conducted two sessions of the TWI Job Instruction (JI) 10-hour class during the first workshop, one class for the Operations Track group and one for the Learning Center Track group with some extra room for local stakeholders to participate. Following the standard TWI delivery format, we did the class for 2 hours a day per class, with five sessions spread out over the whole week, making a total of 10 hours of training time for each group. This gave us a strong starting point because everyone went through the TWI JI class together, and it created that "common language" we were desperately seeking. By giving all members of the team basic knowledge of the training methodology we were to promote, everyone received the same message at the same time, and we were all in synch.

Following the guide, which was the SWC we were putting together, we began the systematic breakdown of the tasks into the TWI Job Breakdown format, going through each step of each operation from start to finish. We took several of the tasks from the pilot area and began practicing this breakdown skill we were learning in the daily TWI sessions. This was one of the most time-consuming tasks because everyone needed to be heard and pitch in, and they did. Every single participant had an opinion about the different operations; surprisingly, it seemed as if any resistance toward sharing disappeared. All aspects of the operations were dissected and analyzed. This was their specialty, and they knew exactly what they were talking about. This work slowly but steadily began to cement the team into one functioning unit.

Videoconference Wrap-Up

At the end of each workshop, we held a videoconference evaluation meeting at which all of the members of the steering team, key stakeholders, and all other relevant stakeholders could join. The team presented the findings and results of the week's work by first defining the overall program and goals for the week followed by what was achieved during the workshop. After the presentation, the stakeholders were encouraged to ask questions in addition to giving their own thoughts and inputs. This created a dynamic atmosphere and made it possible for us to keep a close watch on the direction of the project from the point of view of the stakeholders.

During this first joint process evaluation, it became clear that we needed to develop a process for selecting and evaluating the local job trainers who would be brought on board after the pilot was completed and approved. This assessment was brought to the fore by the molding general manager and the human resources (HR) director in Mexico, who wanted to make sure that the right people would be selected to fill the job trainer role we were creating and testing. The Mexican factory had a clear agenda for this aspect of the pilot project; they knew they had to build the best-possible foundation for the growth that was coming to them soon. In the end, success with this growth would depend on their training resources and how they could make improvement here. It was decided to initiate this as soon as possible, right away during the first "homework" phase and the second upcoming workshop. The work executed during this time period is today considered one of the key factors in the subsequent successful training effort during the production ramp up in Mexico.

During this first workshop, the participants basically followed Tuckman's first stage of group development, "forming," to the letter.* This created some concerns for us on how the group dynamics would evolve during the homework period, video meetings, and at the next workshop, which was to be held in Denmark. After all, following the forming stage comes, as we well knew, the "storming" stage.

Workshop 2: Denmark

The second workshop was scheduled to be held in Denmark in week 26, but before we got that far, it had become clear to us, during the homework period, just how motivated and eager the team was to show off what they were learning. It had also become evident to all of the participants that there

* Bruce Tuckman proposed in 1965 that group development follows four phases: forming, storming, norming, and performing. (Tuckman, Bruce W. 1965. Developmental sequence in small groups, *Psychological Bulletin* 63, 384–399.)

was more to developing the perfect Job Breakdown than might have seemed apparent at first sight.

We touched on this at the end of the last chapter, but one of the more vital issues that needed addressing was that the TWI Job Breakdown Sheets we were creating had become too wordy, and the Important Steps, Key Points, and Reasons for Key Points we put in them were not concrete or clear enough. For example, the Reason for a Key Point like "Wear gloves" could be "Due to safety issues," which is good. But, if you elaborate on it by describing it more concretely and clearly with "Touching hot plastic directly causes severe burns," it will be much better. We needed extra help to sharpen and improve this skill, and Patrick Graupp (one of the authors) came in during this workshop to coach us on making good breakdowns.

The Whole Team: Cultural Training

At this workshop, we went deep into the development of team spirit, objectives, rules, and values. Our core mantra was, "The more we share, the more we get," and this created a strong culture within the team and really boosted the interaction between members and hence the knowledge sharing.

The team had expressed during the first workshop that they would like more guidance and direction on cross-cultural issues and how to work on a global team. So, we had hired an external consultant company to give a 4-hour introduction to the global teamwork approach. During this introduction, the team used cultural profiles they had prepared for themselves before the session based on different cultural dimensions. These profiles not only supplied us with a concrete tool to handle the various confrontations and stumbling blocks that might occur in a global team context, but also gave us a platform to start a dialogue with each other. This ignited a healthy process by which each team member was able to help illuminate unsolved issues.

TWI Introduction to Stakeholders

In an attempt to maintain and gain interest in what we were doing, as well as further convince the stakeholders to support the changes we were proposing, we decided to invite them to a TWI introduction module and a presentation of the global pilot. We planned a 2.5-hour presentation at which Graupp, the TWI senior master trainer who we now had in Billund, Denmark, to do the follow-up course work, would use the first 2 hours to go through parts of the 10-hour TWI JI course that would give a good general overview of the method and its purpose. This was followed by the global project presentation, which covered the plans for the pilot project as well as the new job training organization. After the meeting, we received extremely positive feedback in terms of plain interest in both the instruction method and the project itself. Because the presentation was such a success, we decided to repeat it after each workshop in each plant going forward with the purpose of helping make the ultimate implementation phase much smoother after the pilot project was completed. Today, this has become a standard part of all pilot project workshops dealing with training.

Learning Center Track

During this workshop, the focus for the Learning Center team was to wrap up development of the selection process for job trainers, which was the strong request we received from the management of the Mexican facility at the end of the first workshop. Each consultant had developed and processed a lot at "home." These bits and pieces now had to be put together to form the outline for the selection process. The selection process was tried out during the workshop on two of the global trainers to share insights among the Learning Center consultants and adjusted accordingly. Moreover, it had been decided that the establishment and development of a "soft skills" trainer course (the Training and Learning Skills Course), which could add support to both the trainer role and the TWI JI program, should be launched. This course would take as its starting point the DiSC model* and, in addition, try to link modern learning theories to TWI JI to explain why the JI method works as well as it does. You can read more about this in Chapters 6 and 7.

Operations Track

The planned agenda for this group was the further development of the Job Breakdown process; for the most part, this meant practice—practice,

* Behavioral assessment tool based on four different personality traits: dominance, influence, steadiness and conscientiousness.

practice, and more practice. During the TWI JI follow-up session that was conducted during this week, we finally managed to develop and perfect a LEGO facilitation method for breaking down jobs in the global team (described in the previous chapter). We also learned that making good breakdowns is as much an art as it is a science. It not only was the technical aspects of doing the job that we were learning to discover and record but also finding the right language, few words and simple, and this was the real skill of the technique. We were all technically oriented people, and when Graupp told us that what we were doing now was more the work of poets than engineers, we had to swallow hard and keep practicing.

As predicted by Tuckman's model, after the forming stage comes the storming, and we were in a process of dynamic cultural change. This became one of the key issues for all the following workshops, along with creating a commonly understood and agreed-on terminology. There were several times when our team members had heated exchanges, but these were carefully resolved using the cross-cultural tools we were delivering to the members, resulting in a higher level of cultural awareness and understanding. These were the first steps in creating the ideal structure for the Global LEGO Training Organization. It is of the utmost importance to understand that this kind of scenario can happen in any organization, and it is not a matter of lack of respect or poor manners. It is only a question of lack of cultural understanding. This is true whether working on a global scale rollout or simply communicating between the day shift and the night shift. It goes for all levels and areas of the organizational structure.

Workshop 3: Hungary

Next, we moved on to Hungary, the location of our first planned workshop that had to be put off because of a volcano eruption. Even though the main focus here was, for the most part, still placed on the Job Breakdown process, we needed to put aside time to practice the TWI JI four-step training method. We learned that once you have developed a first draft of a TWI Job Breakdown Sheet, you must test train it to find out if it works. Just because you feel that you have found the right Important Steps and Key Points and put them into words that appear to you to be simple and easy to learn and remember, you can never know for sure until you have someone try to learn the job using the breakdown. Here, you must, as always, use the correct methodology to teach it, which is given in the JI four-step method. So, road testing our breakdowns gave us ample opportunity to practice and hone our skill at using the training method.

Coaching the TWI Job Instruction Four-Step Method

We quickly discovered that practicing the JI four-step method was a necessity for it to become a natural part of regular day-to-day training. This was especially true for the job trainers who had been trainers previous to the pilot project as they had to unlearn a lot of old habits to grasp and internalize the new method. This is why we turned to starting each session by developing a rough draft of a TWI Job Breakdown and then test training it three to five times to practice the four-step training method, making small adjustments in the TWI Job Breakdown Sheet as we did this. This was also a great opportunity for the trainers to help each other gain insight into the different job methods they were breaking down separately by training one another.

During these practices, we always tried to pair two job trainers with one Learning Center consultant. In this way, the Learning Center consultant would be able to observe and give immediate feedback to the trainer who had performed the training. As an aid to this task, the Learning Center consultants developed a structured evaluation sheet of the four-step training method. This helped generate useful feedback on the use of the method and gave attention to the softer aspects of the training. (See Figure 5.2 for an example of this evaluation form.)

After this procedure had been completed three to five times, the trainer would assume the Learning Center consultants' observer position, which helped the trainer develop skill in giving feedback in addition to just conducting the method. Because giving feedback is such an essential part of the trainer's job and can be a hard skill to achieve, we also focused on this in our internal training and learning course.

Having the trainers evaluated training a job they had made a breakdown for before sending them off to start training employees not only ensured that they were well equipped with the right skill set to train but also gave them a confirmation that they were indeed ready to take on the assignment. During the pilot project, we found that it took 3 to 4 months before trainers really started to understand the method and perfected their technique at using JI, a method that would help them handle even

Job Instruction Training Evaluation Guide

Trainer:	Job:	Date:

Preparation for training

Yes/No — Comments:

- [ ] Job breakdown sheet complete?
- [ ] Work area was set up for training?
- [ ] Neat and orderly?
- [ ] Tools and equipment ready?
- [ ] Safety?

Step 1: Prepare the student

Yes/No — Comments:

- [ ] Put the student at ease?
- [ ] Told them the job name?
- [ ] Found out what they already know?
- [ ] Got the student interested in learning the job?
- [ ] Placed the student in the correct position to learn?

Step 2: Present the operation

Yes/No — Comments:

- [ ] Demonstrated the job one major step at a time?
- [ ] Repeated the operation stressing the key points?
- [ ] Repeated the key points explaining the reasons?
- [ ] Instructed clearly, completely and patiently?
- [ ] Did not give more than student could master at one time?

Step 3: Try out performance

Yes/No — Comments:

- [ ] Had the student do the job while correcting errors?
- [ ] Had the student explain the major steps?
- [ ] Did the job again while explaining the key points?
- [ ] Had the student explain the reasons?
- [ ] Continued until the student understood the job?

Step 4: Follow-up

Yes/No — Comments:

- [ ] Assigned the person a specific task?
- [ ] Identified who to see for help?
- [ ] Indicated when they would check back?
- [ ] Encouraged questions?

Process check

Did the student adequately learn task?

Was there any confusion?

Did the student struggle with any part of the task?

Ask the trainer what they would do to improve the training?

How will they incorporate these corrections into the follow-up?

Figure 5.2 Job instruction four-step method evaluation form.

the most stubborn "students" and create sincere motivation among the workforce to do the work following the standardized method. To develop yourself into a strong TWI job trainer is like learning how to play the piano: You cannot learn it by reading a book—you need to practice it to master the skill.

The Whole Team

The whole team worked with one of the local Lean consultants from the LEGO Group's Danish Lean Office to create a relationship between the TWI Job Breakdown Sheet and the Work Element Sheet. The Job Breakdown Sheet would support the trainers during the training process and help generate a structured image of a given job, which would help simplify the training process. The Job Breakdown Sheet was then developed into the Work Element Sheet, which is supported with images and other details and was to be used by the workforce and first-line leaders after the training as a reference document (see Figure 5.3).

We introduced this improvement to the training strategy at this workshop. There was a general consensus that linking the two tools made perfect sense. The Job Breakdown Sheet was designed as a trainer tool and therefore had no real value for operators or first-line leaders after the training except perhaps as a reminder of what was learned in the training when, for example, the job had not been done for a considerable amount of time. Because the breakdown sheet was designed and created for the trainer's use, it would not make much sense to those other than the trainers.

The Work Element Sheet, on the other hand, was a valuable asset to operators and first-line leaders precisely because it did make sense to them; here, the Job Breakdown Sheet was backed up with more text as well as images of, at a minimum, all the Key Points. It also supplied the first-line leaders with a document they can use for process confirmation, including time study, which was added later. The SWC (job overview) and Work Element Sheet (job content in detail) in combination provided operators and first-line leaders with greater insight into the job procedure *after* the training had been completed. The operators were then encouraged to study these documents before seeking any needed help from the trainers.

Learning Center Track

It was essential that the TWI JI Trainer Manual, the JI Pocket Cards, Job Breakdown Sheets, JI Participant Guides, and other materials that came with the TWI course be translated from English into the local languages that would eventually be used for teaching. This needed to be done to ensure complete understanding of the methods at all levels of LEGO Operations so that, no matter where job instruction was given, any member of the LEGO group would be trained in exactly the same way. Finalizing the translations of the TWI JI material

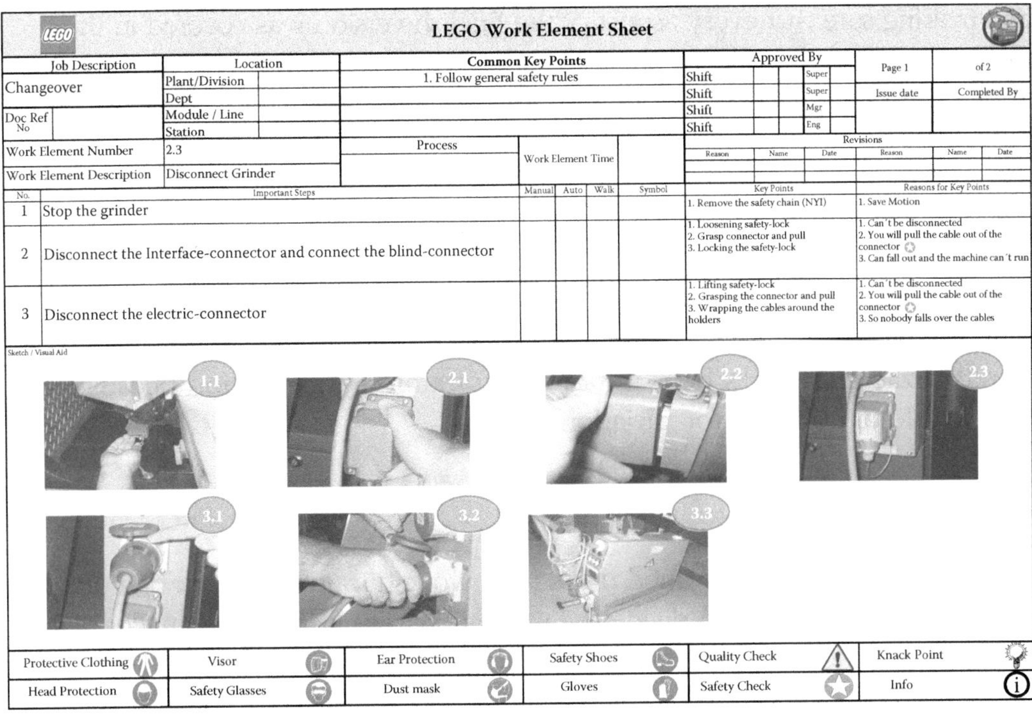

LEGO Work Element Sheet

Job Description	Location		Common Key Points	Approved By		Page 1	of 2
Changeover	Plant/Division		1. Follow general safety rules	Shift	Super		
	Dept			Shift	Super	Issue date	Completed By
Doc Ref No	Module / Line			Shift	Mgr		
	Station			Shift	Eng		
Work Element Number	2.3		Process	Work Element Time		Revisions	
Work Element Description	Disconnect Grinder						

No.	Important Steps	Manual	Auto	Walk	Symbol	Key Points	Reasons for Key Points
1	Stop the grinder					1. Remove the safety chain (NYI)	1. Save Motion
2	Disconnect the Interface-connector and connect the blind-connector					1. Loosening safety-lock 2. Grasp connector and pull 3. Locking the safety-lock	1. Can't be disconnected 2. You will pull the cable out of the connector 3. Can fall out and the machine can't run
3	Disconnect the electric-connector					1. Lifting safety-lock 2. Grasping the connector and pull 3. Wrapping the cables around the holders	1. Can't be disconnected 2. You will pull the cable out of the connector 3. So nobody falls over the cables

Sketch / Visual Aid

1.1 2.1 2.2 2.3 3.1 3.2 3.3

Protective Clothing	Visor	Ear Protection	Safety Shoes	Quality Check	Knack Point
Head Protection	Safety Glasses	Dust mask	Gloves	Safety Check	Info

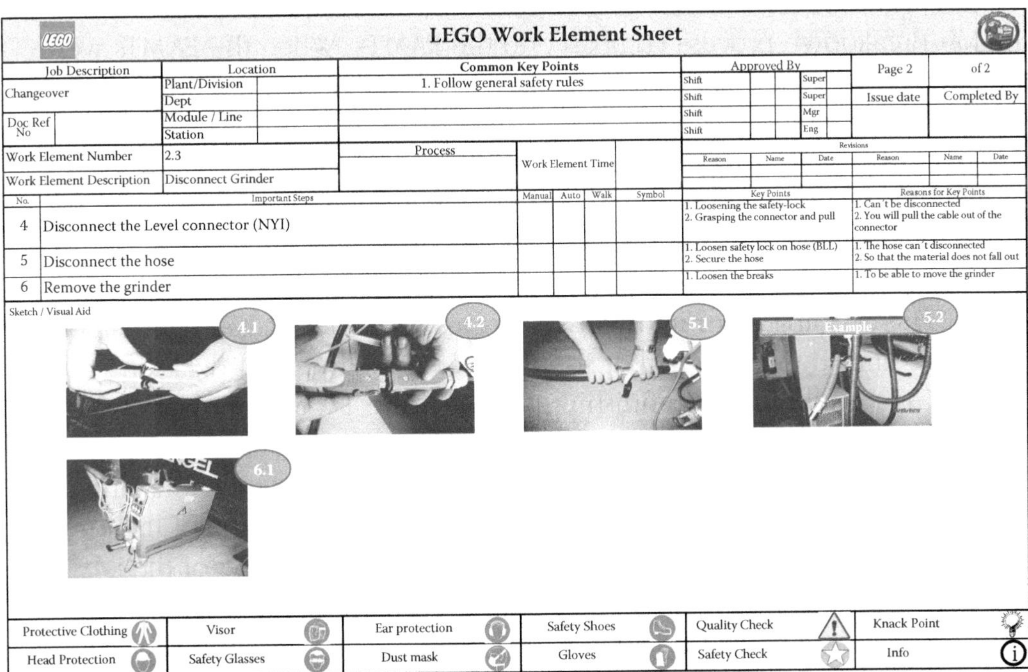

LEGO Work Element Sheet

Job Description	Location		Common Key Points	Approved By		Page 2	of 2
Changeover	Plant/Division		1. Follow general safety rules	Shift	Super		
	Dept			Shift	Super	Issue date	Completed By
Doc Ref No	Module / Line			Shift	Mgr		
	Station			Shift	Eng		
Work Element Number	2.3		Process	Work Element Time		Revisions	
Work Element Description	Disconnect Grinder						

No.	Important Steps	Manual	Auto	Walk	Symbol	Key Points	Reasons for Key Points
4	Disconnect the Level connector (NYI)					1. Loosening the safety-lock 2. Grasping the connector and pull	1. Can't be disconnected 2. You will pull the cable out of the connector
5	Disconnect the hose					1. Loosen safety lock on hose (BLL) 2. Secure the hose	1. The hose can't disconnected 2. So that the material does not fall out
6	Remove the grinder					1. Loosen the breaks	1. To be able to move the grinder

Sketch / Visual Aid

4.1 4.2 5.1 5.2 Example 6.1

Protective Clothing	Visor	Ear protection	Safety Shoes	Quality Check	Knack Point
Head Protection	Safety Glasses	Dust mask	Gloves	Safety Check	Info

Figure 5.3 Example of Work Element Sheet.

and making sure that every aspect of the English version was covered in the local manuals was a topic for the Learning Center team during this workshop.

We put much effort into following the TWI Institute's guidelines, which have truly shown their worth. The fact that we translated and used the guidelines 100% and to the letter is the reason for our ability to train and teach exactly the same material across the plants as the training organization evolved. One of these outcomes can be found in the development of a common terminology for the work processes as this created a common structure and identity for the whole global organization. This is what determines work standardization and lies beneath a healthy training organization because it leaves little to no room for variation.

Operations Track

From the experience we gained during the workshop in Denmark on cultural differences, we decided to divide the Operations Track into two smaller teams, one with just the functional area master trainers (FAMTs) and the other with the global job trainers (GJTs). The primary reason for this division was that we had seen occurrences when the GJTs were more reluctant to participate in the Job Breakdown process compared to the FAMTs. When the FAMTs were absent from these sessions, they appeared to speak more freely and express a more personal attitude toward the work. So, apparently, in spite of our effort to create an open environment where all contributions were valued equally and all opinions would carry the same weight, the official hierarchy had rubbed off on the team. Another, and more practical, reason for the division was that there were assignments more suitable for the GJTs and others more suitable for the FAMTs. Dividing them into groups was also a matter of getting the most out of the limited time we had. It soon proved to be the right decision.

The GJTs would focus on finishing the SWC, the TWI Job Breakdowns, and the Work Element Sheets for the pilot case. With the FAMTs not now present in these sessions, the communication became more dynamic and inspiring. In addition to this, the GJTs introduced the idea of getting some of the local employees or stakeholders involved to broaden the team's view of the jobs and gain insights from their expertise. The FAMTs, on the other hand, turned to the Learning Center consultants to start establishing a baseline and a future platform for basic knowledge training by developing a competence overview (see Chapter 7 for more about this). The idea behind this work was to list the jobs the mold setters were doing and then develop a plan to carry out the training. During this process, we learned that relying on only one training approach would be incomplete, which is why we decided to have a blended learning approach in which the development of a new mold setter would be a mix of classroom training, TWI JI training, and apprenticeship.

The FAMTs also spent a great deal of time refining a *collective development plan* for the level 1 mold setters (there are four levels). By doing so, we took the first step toward creating a collective career path for all setters of LEGO Operations in the future. Having the same process performed identically across borders is a huge benefit when you are dealing with globalization. This provides a clear chain of command and helps ease communication. Even though this was not part of the original scope for the pilot, we saw it as extremely beneficial, and it was something that many people throughout the organization wanted to discuss when they met with us. We immediately saw it as a bonus to the project.

All of the great work that this new part of the Operations Track team spawned was, of course, most welcome and positive. It did, however, add extra pressure to the driver of the Operations Track as there now were two teams to deal with, and both needed a great amount of facilitation. This proved an extremely demanding task, and we barely made it through the workshop. This was an issue that needed to be dealt with during the next workshop, which was scheduled for Denmark.

Workshop 4: Denmark

In Workshop 4, we were back in Denmark. During the regularly scheduled Monday morning introduction, we learned a lot from the presentations made by the molding general manager, HR director, and Lean director of the Denmark facility. These introductions were starting to become excellent ways of kicking off the week, and it was almost as if the sites were competing against each other to create a good impression for the global team. This generated good motivation for the team and showed us that the project was being noticed.

A consultant from the local Lean Office had arranged a presentation for us on the process confirmation procedure. Process confirmation would function as insurance that the people trained would maintain their production methods and output after they had received the training. The primary reason behind the presentation was to create a clear impression and understanding of how process confirmation would be used in practice after it was adopted by the production leaders.

The second tool the Lean Office introduced to the team members during the workshop was the skills matrix, and this aligned perfectly with the TWI Training Timetable, which was actually that week's focus in the TWI training portion of the workshop. The Lean consultant covering this topic for us split his own work 50% as a Lean consultant and 50% as a Learning Center consultant; this mix was a good starting point for success in teaching us this material.

The Whole Team

One of the clear benefits in the pilot project of having a close collaboration between the Learning Center and Operations was that the whole team was able to test a newly developed method in a realistic, global environment immediately after it was developed. A good example of just how this benefit was conveyed occurred during this workshop, at which the Learning Center consultants had prepared a new training module concerning feedback and taking an appreciative approach to learners. The reason for developing this training was that trainers often find themselves in situations in which they have to give feedback to or motivate workers, and we saw this as a skill that needed improvement. The Learning Center group needed to test their new material on the Operations Track team to make sure the content was valid and ready to be implemented in the Training and Learning Skills Course they were developing. The feedback from these modules was good, and the material was adjusted with some minor changes and then translated and implemented into the course.

As the pilot project evolved and grew, our mantra, "The more we share, the more we get," became increasingly essential for the whole operation. The project had progressed and so had the members of the project team; they were now eager to share and discuss their various viewpoints, suggestions for improvements, and documents they were making. On the other hand, it became increasingly difficult for us to handle all of the information we were generating, and we saw the necessity for devising some type of a solution to this problem. It was the newest member of the Global Learning Center, an intern from Aalborg University in Denmark, who solved our problem by developing and setting up a SharePoint site on the network for everyone to use. SharePoint is a standard part of the Microsoft Office Suite that allows numerous users to store, share, and correct documents as well as work with common usage tools like calendars and forms. In addition to this, he also took charge of introducing and teaching the new tool to the members. He became so dedicated to the project that he ended up helping facilitate the GJT group, which had become a problem during the last workshop when they split off into a separate group to work on unique assignments, and he did this throughout the final workshops as well as on their video meetings during the homework periods.

Sudden Realization

In the midst of all of this, the FAMT who represented the Hungarian plant and was one of the first involved in the project suddenly had an epiphany and realized the magnitude of the program he was helping to create. He was stunned when it finally hit him how much the local implementation relied on him and his GJTs. Not only did all the GJTs have to prepare and train the future local job trainers, who would in turn train the local employees on the jobs, but also they had to make it clear to middle-level management just how important training was to the whole organization. They would be responsible for a certain amount of marketing, so to speak. All of this would be required for the startup of future projects in the rollout phase after the pilot project closed.

The underlying reason for the pilot project was to convince management of the value and necessity for a good and reliable training standard. The Hungarian FAMT's main concern revolved around the implementation of the training strategy, which made him think, "If it has taken me 6 months to come to this realization, how far along were the other FAMTs in their development?" And this was just half of it. If it took 6 months for individuals like himself who basically "breathe" training and were this close to the process to come to this realization, how long would it take for the local job trainers; the first-, second-, and third-line leaders; employees; support functions; and upper management to reach the same point and the same conclusion? One of the biggest fears among the FAMTs was also the implementation itself. They knew that they would have to spend a certain amount of time training the to-be local job trainers, and this meant that there might not be any sign of improvement at the beginning. So, what would happen if management saw this as a sign of failure? Would this then reduce support for the project?

All these continuing worries needed to be dealt with quickly. We decided to have individual sessions with each of the FAMTs and through these sessions create a list of their concerns and doubts. It was a pragmatic approach. If one of the FAMTs felt that he or she lacked the necessary support from management back home, we simply arranged for a meeting at which we asked the relevant stakeholders there how they viewed the prospects of the work of the training organization in their local facilities. This approach quickly cleared many misunderstandings and put the FAMTs at ease.

The bigger benefit we received by jumping immediately on this concern and taking it off the table was that we now knew the FAMTs had taken full ownership of the project.

Learning Center Track

During the second Denmark workshop, the Learning Center consultants were focused on the further development of the Training and Learning Skills Course. During this process, they were especially aware of the content versus the target group and focused on how to make the best use of the designated time. In addition to this, they gave great support to the FAMTs in creating learning objectives for the different roles in the collective development plan they were creating for mold setters.

Operations Track (Global Job Trainers)

The Operations Track was still working on the ongoing process of developing Job Breakdowns and a training plan for the pilot. In particular, the order in which the Job Breakdowns should be taught or presented to the employees was a topic of concern. Should they be presented in chronological order, or was it actually better to ease them into it by starting with the less-complicated ones first and working up to the more complicated ones? The different sites tried different solutions for this, but we found out the best way was to start with the easy ones. During this workshop, it also became clear to us just how much better the GJTs had become in developing the Job Breakdowns.

By practicing the TWI four-step method repeatedly, they had reached a new level of understanding of the Job Breakdown tool; this was evident in the quality and the efficiency by which they were now developing the breakdowns. This proved unequivocally to us that developing strong and sustainable Job Breakdowns is a skill that takes time to acquire. It is one thing to understand the concept and structure of the breakdown and to be capable of executing the mechanical part of the job, but it is something completely different to deliver a unique and empathic training in which the trainer is adaptable and the training is immersive. This demands a great deal of repetition and insight into the softer side of the trainer's role; it requires adaptability to one's fellow employees depending on their individual backgrounds and aptitudes and, through this, generates the desired stable processes for which safety and quality come first and the job is carried out the same way and with the same tack each and every time it is performed.

To undergo the change from being a solid and dependable coworker to becoming a great trainer takes time. In our experience, it took from 4 to 6 months working with training 2 to 3 days per week. In addition, we believe that it is an ongoing process of learning and evolving one's skills, but you can only get better.

This was indeed a workshop that was characterized by realizations and progress. Besides the GJTs taking a quantum leap in their training skill and mind-set, the project had progressed so much that we felt it was now suitable to present our work to and be tested by the various stakeholders. This is what we defined as the "test of Job Breakdowns."

Testing the Job Breakdown with Relevant Stakeholders

The trainers had established solid Job Breakdown Sheets, and it was time to invite the relevant stakeholders over for a presentation of their findings. This would not be your standard PowerPoint presentation, where you show your findings in neatly made slides, but an actual demonstration of what a real training situation looks like. The stakeholders in our specific case included production leaders, specialists from quality assurance and safety, Lean consultants, and members from the PTD (Production Technical Department). You might think this was just another presentation, but it is important to bear in mind that the trainers were under a lot of pressure in front of this group. Many emotions were riding on a meeting like this: "Have we performed well enough?" or "Will all of our hard work go to waste?" This is why it was essential to prepare the stakeholders beforehand, making sure they were aware of their role and ensuring they would not get lost in minor details. Also, it was important to have them understand that their role was to encourage and give professional feedback.

Job Breakdown Sheets were handed out to all the stakeholders participating in the presentation. The sheets consisted of notes for training and were written in a style that assisted trainers to teach jobs in a way that workers on the floor could easily understand and decipher. This was an important fact to emphasize to engineers and other technically trained stakeholders so that they would focus on the relevant training issues at hand and not start "nitpicking" at whether the right terminology was used. In the long run, relevant stakeholders should really participate in the full TWI JI 10-hour course as this will supply them with the necessary understanding of the importance of the TWI training method, the *how* and the *why* of the Job Breakdown process, and the methodology of good training, which they may not fully grasp.

While it was essential to emphasize to the stakeholders that their input was valuable, they were told not to interrupt until the demonstration was over. The input given by the stakeholders was to be respected, which is why the trainers would receive and acknowledge these with nothing more than a simple "Thank-you." This rule was enforced to have everyone keep an open mind and sustain the need to make a group evaluation. For example, after the demonstration, if one trainer suddenly blurted out "great idea" or "that makes no sense" in response to a suggestion from the stakeholders, the trainer in question would be held responsible for that statement. In the end, though, this was a group process, and whether the statement was positive or negative, it needed to be up to the group to decide whether the suggestion should be implemented to the Job Breakdown Sheet and hence become part of the job instruction.

The input from the stakeholders was in fact important to the trainers, and if they wanted to continue receiving their expertise in the future, professional and timely feedback from our group was also needed. Therefore, part of the presentation process was for the trainers to go through the input they received from the stakeholders and respond the next day. This could vary from a big, "Thank you, your feedback was great," to "Thank you for your input, but we won't be taking that course." Explaining the why to the stakeholders was also important, such as why their input was or was not included in the Job Breakdown Sheet.

During these presentations, one of our trainers would be in charge of demonstrating the training process to a real operator to give a clear image of what the JI process was. Following this, the stakeholders and other observers could give good suggestions and thereby improve the Job Breakdown further. There were two reasons for doing it this way. First, the trainers might have missed or misunderstood an Important Step or a Key Point concerning safety or quality, and this is why specialists in these disciplines were needed at the presentations. The second point, which might seem a bit "soft," was for the stakeholders and observers actually to praise and congratulate the trainers on their JI and the process they went through to provide it. It might seem a bit superfluous, but this is what gave the trainers the necessary confidence to carry on with the instruction and take it all the way to the finish line.

In addition to these two points, there was a third reason for the presentations; this concerned the production leaders, the ones actually in charge of the workforce. It was essential for them also to understand this new work procedure and how it would be trained so they would be able to inform the workforce about the upcoming training and support the use of the new method. This was not necessarily obvious to all. In certain situations, production leaders are reluctant to train and will not appreciate the value of a well-trained workforce. Furthermore, having the production leaders on your side is a definite bonus as they are the ones who prepare the workforce for the training. There will also be workers who are frightened of the change or perhaps see training as a possibility for management to downsize the workforce. So, it is best that they are given the reasons for the training by someone they trust, their leaders.

Workshop 5: Mexico

Just before the fifth workshop in Mexico, all of the GJTs tested the new changeover process at home and recorded it on video to compare it to the first time they had performed the procedure many months earlier at the first workshop in Mexico. The progress from the first workshop to that point was huge. In the first workshop, the variation in time between the different plants doing the same process was 90 minutes. Now, in Workshop 5, this variation time was reduced to a mere 2 minutes. During the first workshop, there was one site that had skipped measuring the mass temperature to save time. That had now become standard procedure for every plant, and everyone knew why they had to do it. We could look at the three videos side by side, and they all looked the same. There were, of course, some small local differences in equipment, tools, and layout, but if they could do the same things, they did. This degree of global alignment in the team showed strong potential.

The Whole Team

The primary objective for this workshop was to finish all assignments connected to the local implementation process. The FAMTs and the Learning Center consultants started by developing a draft for a possible A3, a planning tool pioneered at Toyota, and TIP (Tactical Implementation Plan) for the local implementation of the knowledge we were consolidating during the pilot project. The process of developing A3 and TIP was a real eye-opener for understanding how different the local needs and points of view actually were. This proved to be a demanding task because the local needs have to be fulfilled but, at the same time, have a clear link to the global teamwork. This meant that the local TIP would contain both how to train locally and identifying what the next job would be for performing a TWI Job Breakdown. This could be anything from color change to operator maintenance, all of which should be assigned to a local site to develop and thereafter shared globally with all the other sites.

To this point, it had been the Learning Centers' responsibility to develop the A3 and TIP for the pilot project, but now that we were beginning to think beyond the pilot to the full rollout of the new training organization, we were confirming who would manage the process on a local basis going forward. From the start, it had been planned for the FAMTs to manage the local process, so it was no secret that this workload sooner or later would be placed on their shoulders. When we finally began planning for these things, though, the overwhelming assignment load surprised them to some extent. But, they managed to take control of all the responsibilities and carried them forward to their completion.

The objective was now to present the A3 and TIP to the local molding management and, based on their feedback, create a local steering group that could help carry out the local implementation. The real insight here was that the local general manager of molding, who had been part of the steering team for the pilot project, was now the obvious choice to be included in the local steering group.

Learning Center Track

In the Learning Center Track, there were three major issues that needed to be resolved as quickly as possible: finalization of the Training and Learning Skills Course, the trainer evaluation process, and a recruiting and development plan for new Learning Center consultants.

First was the finalization of the Training and Learning Skills Course. This was the course that would help build the training organization by preparing trainers before they went out to train the workforce. This organization (explained in Chapter 6) would consist of new trainers who would be recruited to perform the training of local workers. Use of the course would begin immediately after the completion of the workshops, so there was high pressure to have it done in time. The course needed to be ready when the local job trainers were selected.

The second issue, which was closely related to both the Training and Learning Skills Course and the skills taught in the TWI JI class, was the evaluation process of the job trainers' progress within these skill sets. We needed confirmation of their knowledge and how they approached their new role as job trainers. To do this, we would have to look at how they used all facets of the TWI JI method, how they prepared and executed training, and how they used their new skill set when they executed training. We knew that the skill set the trainers possessed was something that needed constant evaluation for it not to reach a standstill or, even worse, deterioration. This is why we added quarterly evaluation sessions to the evaluation plan to follow up on their professional progress.

Besides the more obvious reasons for the evaluation process, there was a subagenda to this action. We saw the evaluation of the new trainers not only as a confirmation that they were prepared to undertake the task at hand but also as a confidence-building exercise to affirm this in the trainers themselves by acknowledging their skills and the months of hard work they had dedicated to the effort.

The third issue that needed to be solved was the plan for the recruiting and development of new Learning Center consultants. As we were closing in on the final stages of the pilot project, the need for more consultants would increase with the launch of several new projects in the rollout phase (see Chapter 8).

Operations Track (Global Job Trainers)

One of the most discussed and worked on subjects during this workshop was how to use the pilot project as a catalyst for implementing the Job Breakdown approach to other changeover processes in the molding department. As molding of the LEGO bricks and minifigures is the core of the LEGO Group's work, this was a huge mission with 32 different types of machines, 8 types of casts/molds, and 8 types of plastic to be processed, all of which needed to be covered. We had been using the method in the pilot on one type of machine, one type of cast/mold, and one type of plastic. Would it be possible to transfer our work to other types of machinery and material? This is where all the work we had performed paid off. By the end of this workshop, using the Job Breakdowns we had already made, we found that we would actually be able to cover training for up to 80% of all the changeovers within molding.

Operations Track (Functional Area Master Trainers)

The FAMTs' nonstop work, together with the Learning Center consultants, on the basic knowledge theory and the development plan for level 1 mold setters was starting to take its toll on all involved. Many of the FAMTs were growing frustrated with the plans as they were not completely identical between the sites. The idea of a global rollout of the learning organization was for it to be transparent and unified, was it not? This went for the job descriptions as well, which they still had not yet developed for hourly paid workers. This is why it was decided to finish with these issues during the next workshop, or so they thought.

Workshop 6: Hungary

The sixth workshop was cancelled. The majority of the steering team felt that the concrete goals of the project were accomplished or close to it at least. This meant that there was no real business case to validate the need for another workshop as scheduled. The SWC, the Job Breakdown Sheets, and the Work Element Sheets from the pilot project were all complete; the GJTs had practiced and trained all of the Job Breakdowns; each site had completed changeovers on the pilot case; the variation on the pilot case was reduced from 90 minutes to 2 minutes; and there was a clear consensus on the hows and the whys for the process. The FAMTs were getting ready to implement the final touches to the first draft of a global version of basic knowledge training material and a training plan for the level 1 mold setters. The Learning Center consultants had finished creating the selection process for local job trainers, the Training and Learning Skills Course was complete, and the trainer evaluation process had been rehearsed and translated. So, all of the practical aspects of the pilot project were finished.

To continue working with some concerns among the participants on the local implementation and be better prepared for the local implementation, the last workshop would make sense to complete. But we had to move on.

Approval of the Pilot Project

We knew all along that our concept was not perfect, and that TWI by itself would not save the day for the LEGO Group in the transfer of knowledge and skills. But, we did know that it was a strong and well-tested concept, and we also knew that it was the best-possible solution for the LEGO Group to regain strength and structure for building and evolving. Needless to say, we still had some concerns about the next steering team meeting, where we would present the results and recommendations from the pilot project. With the positive indications the pilot project had given us, though, we were ready for it.

As we arrived at the steering team meeting, we sensed, to our pleasant surprise, an optimistic atmosphere. We were all gathered in Denmark for this last meeting, which was a great way to end the pilot project. And, in the end, the project was ultimately approved. During the presentation of the processes and the results, two observations were made by members of the steering team, members who all along had been closely involved with the pilot project and the team itself. The first observation was that this approach to working globally was seen as a "role model" for other projects in the company, and it was definitely something whose use should be encouraged. The second observation was that the TWI JI methodology would allow us to reap the benefit of learning from the employees' tacit know-how, which heretofore had proven to be difficult to capture.

The steering team was also able to understand that there was still a long way for us to go because integrating the Global LEGO Training Organization to the existing LEGO Group systems was easier said than done. The following are some of the issues debated during this steering team meeting along with some of the challenges the team wanted us to dig into during the rollout phase:

- When would we be able to convert the local job trainers into team leaders? It was seen in Hungary that the job trainers had grown with the assignment, and that, over time, they were able to take on more responsibilities.
- There was a need for an in-depth clarification of the roles and the responsibilities between the various support departments: Lean Office, PTA, and HR. We needed a clearer structure of who does what and who you should turn to for support in any given situation.
- There needed to be established a stronger link between the departments working on the development of SOPs, the quality handbook, and the new process tool Aris.
- Training and evaluation of line leaders in relation to TWI and its link to Lean were missing, creating noise in both the TWI and Lean implementations.

After the meeting, we asked each of the team members questions, and some of their answers are given next:

Why did you approve the pilot project?

The project was approved due to the very good results we saw from the "test" of making the training for changeovers and due to the very good process of making this a global training method.

Senior Director, Molding Design and Implementation

If we don't follow the same standards, we don't have a valid baseline we can continuously improve on, which is one of the big building blocks in our LPS [LEGO Production System] journey. Seeing the motivation in the global training team and how they globally aligned the processes—this we could not have done with the old training setup; now they have clear roles and methods that support this. The global training team improved their own performance on the changeover process over 18 weeks in relation to clear safety/quality key points and time variation from 100 minutes variation on the changeover to only 2 minutes variation—this showed a strong potential.

General Manager, Hungary

LOM [LEGO Operations Mexico] had a big task to onboard a total new organization where training was needed in all areas. To support this and to ensure achievement of all our KPIs, we needed standards which only could be done by a global trainer team and a dedicated trained local trainer team.

The global and local trainer teams make local and global networking easier and also focused which again was and are essential for both building a new organization but also for improving performance in all areas.

General Manager, Molding Mexico

We didn't approve a clearly described business case. What we did was approve a strategic direction for HOW to create standards and HOW to do training in production, and I see it as another brick in the LPS [LEGO Production System] journey we started already in 2008. So I approved it because I believed the direction was right!

General Manager, Molding Denmark

What benefits can you now see from the job training concept/organization (financial or nonfinancial)?

Our production processes are more predictable and transparent, now that we have developed clear standards. We have a more clear training plan and the quality of the training is improved, which also has developed clearer roles and responsibilities. Looking at our improvements in safety and process quality, the standard training and work standards have been a big contributor—now we need to ensure that the leaders' role in standardized work is aligned, developed and implemented.

General Manager, Hungary

The global and local trainer teams make local and global networking easier and also focused, which again was and are essential for both building a new organization but also for improving performance in all areas. Without the trainer team it will be more difficult to deploy continuous improvement. In my book, that is the only way to do it and, at least on our site, I see a lot ahead of us in that area. We will still be in an onboarding mode, where the trainer team is a must. Then we have the continuous revisit/refresh of the basic training where also the trainer team is essential. Then we, of course, also have all the TWI tasks, etc.

General Manager, Molding Mexico

The company has been able to train on a big scale in LOM [LEGO Operations Mexico]. We will, as we learn from this process, gain on safety, quality and cost. By improving changeovers, we believe that our target of reducing claims by 30% is positively impacted by the standards and training. The result YTD [year to date] is a reduction of 26%. Part of this is coming from improvements in the way we execute changeovers.

General Manager, Molding Denmark

What effect would it have if we do not implement the training organization?

We lose the global way of training and thereby have to use extra resources (local or from other sites) every time we have to train new employees, and we can't be sure that they are trained in the right LEGO way, meaning we'll have all the thinkable "bad habits" carried on. We lose the new way of global alignment on training and methods—back to the LEGO way from 2005/2006!

Senior Director, Molding Design and Implementation

Without standard training, we cannot sustain our standardized work, meaning that our production variation will increase in safety, quality and delivery resulting in poor performance in our business KPIs. Without standard training, our training/on-boarding time will increase and thereby not support the strategy of improving our flexibility in the packing areas where we want to utilize temporary and current workers better. Without the training organization, it will be a big challenge for us to control quality on-boarding in our to-be molding set up or just to think about the challenges we will meet in building up operations in Asia without standardized training or work.

General Manager, Hungary

I see two major issues if we don't have the training organization:

- We will lose this important way of gaining learning in the organization. How will we continue the LPS journey if we do not dig into the "devil" of the details.
- The on-boarding task will be less efficient and hard to repeat for the next growth in supply chain.

Senior Advisor, Molding Processes, Denmark

From Global Pilot Project to Local Anchoring

After the completion of the pilot project, there were two new major tasks that needed to be accomplished. The first was to initiate the rollout of the new systems to the local divisions based on the experiences gained from the pilot project. The second was to follow up and support the local anchoring of the pilot case at the three plants. It was evident that there were diverse needs within the three local sites. This is why, ultimately, the implementation of the training organization into the local sites was characterized by big differences in speed and approach. This section looks at the return of the FAMTs, the GJTs, and the Learning Center consultants to their home plants and how the implementation of the learned material unfolded there.

Different Needs

The first step in the process of having local management acquire control and ownership of the implementation was to adjust the A3 and TIP drafts, which the FAMTs had prepared, to reflect input from the local steering groups. Looking at how the three sites prioritized the speed and breadth of the changeover process implementation, it was clear that not all sites were equally willing or ready for the change. The plant in Mexico was highly motivated for the implementation because individuals there knew that they were going into a long period of bringing in new employees. The sooner it became a natural part of their operations, the better it would be for them. Those in Hungary had a similar viewpoint because they were informed there would be an expansion of an additional, larger molding plant in their facility. In addition to this, their site had shown excessive variation in their production procedures, which needed to be addressed. This need for expansion proved a great motivator for both Hungary and Mexico. In Denmark, the situation was a little different. There was no promise of expansion, and the focus on cutting costs made them a little reluctant to start the implementation of the strategy.

Despite the differences in motivation, all of the plants began the process of selecting, training, and evaluating local job trainers. Simultaneously, one of the most important jobs was for the GJTs, in collaboration with the newly selected local job trainers, to review the TWI Job Breakdowns that were developed during the pilot project. This process proved to be extremely valuable as it resulted, based on the 30 Job Breakdowns developed during the pilot, in over 100 suggestions for improvement. Even though the pilot project team had spent the better part of 6 months creating these breakdown sheets as "perfectly" as they could, once the operators in their home plants began being trained by them, they immediately began seeing things that could be improved. It can be said that all of these ideas did not just materialize at the time of the training. They had always been there.

But, the new instruction method, based on the TWI Job Breakdowns, gave them a fresh format to look at the contents of the work and a clear medium to articulate and declare their improvement suggestions.

These suggestions were then discussed and compared at the weekly GJT videoconferences. In addition to this, the FAMTs would have a separate videoconference at which the final approval of any adjustments to the TWI Job Breakdowns would take place. Also, the progress of the local implementation would be assessed at that time and the need for any further development of the strategy and basic knowledge evaluated. During this initial phase of the rollout, there was still a unity of purpose to the strategy of the learning organization, but the rollout of the training to the workforce within the local sites would differ to a degree.

Molding Hungary

In Hungary, it was quickly decided to commit the trainers to 100% of their work time on training for certain periods of time. This was done to improve the level of the mold setters as quickly as possible because of the high variation in production that was causing problems when it came to quality control. This approach proved successful. Within a short time, the gaps in production were pinpointed and improved. Based on the results from this intensive training, they developed a new plan for bringing new employees on board that ensured they would gain a higher level of process understanding. It was proven here that they could enhance the pace of the development of a level 1 setter from 1 year to 3 months.

> The new local training organization in Nyíregyháza means a world of difference for employees and their performance. Through structured training, where training is based on a global standard, we have been able to follow up on training and employee performance with far more detail.
>
> **Functional Area Master Trainer,**
> **Molding Operations, Hungary**

Molding Mexico

By this point, the factory in Mexico was undergoing a great expansion process and already had a high focus on implementing the directions that came from the pilot project. They were now ready to build up their training organization by using the selection process, TWI Job Instruction, the Training and Learning Skills Course, and trainer evaluation to the fullest. Here, the strategy really proved its worth.

> We could not have ramped up our injection molding factory in Mexico so fast without the Global LEGO Training Organization. It actually was faster than we had expected. The investment in the training organization and concept really paid off in the ramp up period. And now it ensures that we maintain the future development of our workforce in a structured way.
>
> **General Manager, Molding, Mexico**

Molding Denmark

In molding in Denmark, a local training organization was quickly established, ready to train the employees. During this startup phase, there was intensive training, and the trainers showed great motivation to provide good training and good results. The local FAMT's statement shows how this worked:

> The TWI Method we are using is being well received, even by our most experienced employees. They are ready to change their old routines, because they, through the TWI method, get the reasons behind why they are being trained in the different processes. Safety, quality and how our equipment is being handled has also been positively changed.
>
> **Functional Area Master Trainer,**
> **Molding Operations, Denmark**

The local Danish molding management and workforce was, however, not 100% ready for this type of investment and had serious doubts at first about the implementation and the use of the method. It has since, however, gained pace, and this has resulted in some good outcomes for molding in Denmark by using training as a means to reduce variation. For example, a project was completed on the reduction of errors within the scanning of transport boxes that created solid results. There had been a high error rate in the manual scanning of transport boxes that were moving LEGO elements from the molding to the packing area. That meant that the packing line could possibly be set up with incorrect elements. Going from almost 460 scanning errors per year to only 2 in a month was an excellent result. There were, of course, other factors to this success, but one of them was clearly that all the workers doing this job were trained in the process by use of TWI JI. This created a lot of good attention to the new training processes.

General Notes on the Local Implementations

During this local change process, we conducted several surveys and made observations about the pros and cons of the new training structure. Some of these observations are listed next as before and after items:

Before: There was no clear structure or process for training.
After: Today, there is a clear structure or process for training.

Before: No clear plans were available for bringing new hourly employees on board.
After: Today, there are training plans, training materials, and trainers to support the new hourly employees.

Before: There was no clear link between the development of new processes and training.
After: Today, there is a strong link between continuous improvement and the training organization.

Before: The dialogue around competencies development was abstract.
After: Today, it is very down to earth and realistic.

Before: There was little knowledge and few skills in understanding global transfer of knowledge and skills.
After: There is a higher understanding and recognition of these processes.

CREATING THE GLOBAL LEGO TRAINING ORGANIZATION

Chapter 6

Building the Organization

Introduction

Before beginning to establish the new learning organization, we knew that we had to create a comprehensive set of standards up front to obtain the consistency of performance the proposal required. This would include developing organizational charts and job descriptions for each of the new roles in the training organization. Moreover, we felt that with the focus on creating trainers, leaders, and practitioners, based on the Training Within Industry (TWI) model, we wanted to have a strict system of screening potential candidates to select only the best-qualified people to hold these positions. This chapter describes in detail the mechanics of how these systems were conceived and documented with the goal of building and anchoring global and local learning centers. It also shows the details of the selection process, including trainer competencies and a recruitment plan that would be used to staff the training organization.

Roles in the Training Organization

Based on our previous experiences giving training on an international basis, we knew that global collaboration and a common approach would require much preparation and some initial direction setting. In the past, we would typically send our best specialist from Denmark to, for example, our production site in Mexico, and the specialist would spend several weeks there conducting some on-the-job training for a specific job. After returning, though, we realized that the knowledge and skills for the job had not been properly learned and followed. In addition, challenges arose as some of our technical specialists seemed more interested in showing their high degree of knowledge instead of focusing on the learning process for the people receiving the training. It was a challenge to develop and maintain the necessary job knowledge and skills in the local sites because of this lack of structure and discipline in the training. At the same time, we were struggling with language issues and misunderstandings that created problems that had to be handled.

We realized that our best specialists were not necessarily our best trainers. In addition, some of our trainers were full-time trainers, simply traveling around the world conducting training. When returning to their home organizations, it turned out they were no longer up to date on process changes, technology updates, and the like that had occurred during their absence. The result was that these full-time trainers were actually training incorrect and out-of-date procedures.

We had many long discussions around how to design a training organization that would avoid these problems and realized that what we needed was to establish a more solid and well-defined group of trainers at all of our production sites. We designed a training organization consisting of a functional area master trainer (FAMT; for specific functional areas such as molding, packing, warehouse, etc.); global trainers; local trainers; and technical writers (see Figure 6.1). We then clarified each of the different roles in the training

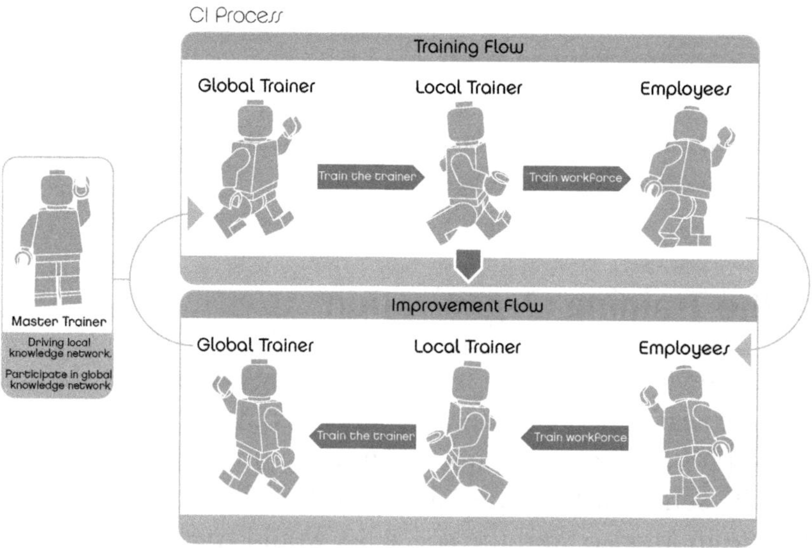

Figure 6.1 Trainer roles.

organization based on job descriptions and Key Performance Indicators (KPIs) for each role, which were then approved by the human resources (HR) directors at all of the production sites.

Being a Trainer Is a Role—Not a Position

We knew from our experience that good trainers had to be continuously updated on technologies, processes, and procedures and had to be part of the production areas they served so they would not lose their functional expertise. We decided then that the different trainer roles would not be full-time training positions; we started by creating a guide to how much of their time the trainers should use on training and how much on their daily operational tasks. In the initial pilot project, we recommended that the FAMTs allocate around 70% of their time to the training role. Later, we realized that this really differs a lot from area to area depending on the number of trainers, the complexity of the training, and other factors. We settled on levels for the global job trainers at around 50% and for the local job trainers at around 30% of their total working time to be devoted to training.

The importance of keeping a good balance between daily operations and training was readily seen when some of the trainers were allocated too few hours for training activities, and they would struggle mastering the training methodology and routine. If, on the other hand, they were allocated too many hours, it would be difficult for them to keep their attention on the work and their daily responsibilities as part of the work area. So, here it was important to find the right balance that created the necessary dynamics between the daily operational tasks and the training activities to ensure sufficient quality and motivation in both areas.

As a rule of thumb, we recommended to our plants that they have 1 job trainer for every 10 employees; this followed recommendations given in *Toyota Talent*, the definitive book on Toyota's use of TWI (J. Liker and D. Meier, McGraw-Hill, New York, 2007, p. 61). We learned to consider this as just a guideline because it will vary depending on the following conditions: speed of implementation, depth of knowledge/skill, size of growth, receiver maturity and skill set, and various cultural dimensions. In addition, it will also depend on the extent to which the trainers have to develop the training material and the standards. The more material to be made, the more resources are required. Table 6.1 sums up our guidelines for the deployment of these different roles.

Table 6.1 Trainer Guidelines

Area	*Master Trainer Guideline: 50–100% of Role*	*Global Job Trainer Guideline: 50% of Role*	*Local Job Trainer Guideline: 30% of Role*	*Technical Writer Guideline: 50–100% of Role*
Specialist area/ function	1 per functional area (master trainers can cover multiple functional areas)	1 per area of expertise (+ scalability requirements[a])		1 per functional area (technical writers can cover multiple functional areas)
Manufacturing	1 per functional area (master trainers can cover multiple functional areas)	1 per 10 local trainers (minimum 2 in each area)	1 per 10 employees (+ scalability requirements[a])	1 per functional area (technical writers can cover multiple functional areas)

[a] Scalability requirements:

- Speed of implementation
- Depth of knowledge/skill
- Individual versus group knowledge/skill
- Size of growth
- Maturity/receiver skill set
- Cultural dimensions

Trainer Roles and Responsibilities

The **functional area master trainer (FAMT)** plays a vital role in leading the trainer teams; there should be one in each production site. He or she is responsible for the following activities:

- Gives content approval for all area training material (Standardized Work Charts, Job Breakdown Sheets, Work Element Sheets, Work Guides, other basic knowledge training material)
- Participates in selection, development, and evaluation of global and local job trainers and is responsible for motivating them in their training duties

- Drives and participates in networking activities within the site and globally within the functional area
- Prepares and runs the daily huddle for job training
- Drives the local trainer organization (matrix organization) in daily activities (e.g., Tactical Implementation Plans [TIPs], progress reports, success stories, KPIs in relation to training, etc.)
- Drives, develops, and maintains the onboarding plans for the functional area roles
- Coordinates with management in relation to training needs and resources
- Develops global TIPs with FAMTs from the other sites within the same functional area

Furthermore, the FAMT closely collaborates with the line management of the functional area. The line manager, of course, is still responsible for human resources development in his or her area and the training of the workforce. Here, these managers are collaborating with the FAMT in relation to startup training and scoping and determining budgets, resources, KPIs for the trainers, and so on. The line management is also measured overall on the performance of their FAMTs and trainers in accordance with requirements in the position profiles. Last, but not least, line managers ensure that the job training process is followed based on the standards, and that there will be proper execution of process confirmation and follow-up in their areas. Figure 6.2 shows the organization chart and how these trainers fit into line management.

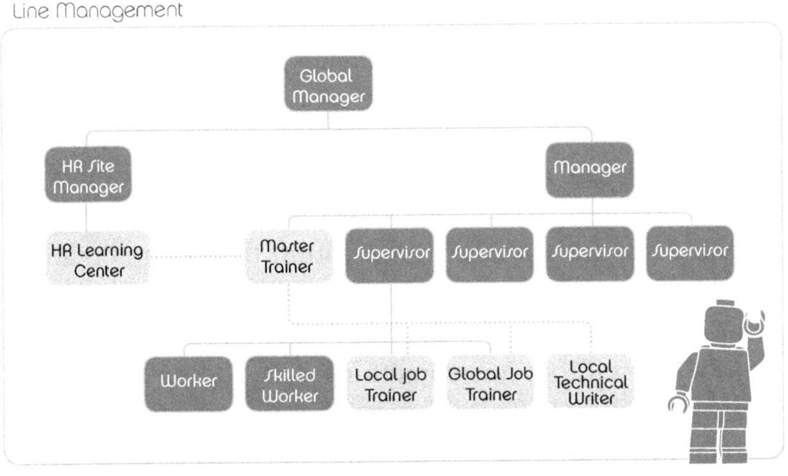

Figure 6.2 Trainer positions in organization chart.

The **global job trainer** (GJT) is responsible for the following activities:

- Prepares, trains, motivates, and evaluates local job trainers, trainees, and vendors (using defined training methods, material, and documentation)
- Follows up on local job trainers and ensures that they follow the global/local work standards with a focus on safety and quality
- Participates in networking activities locally and globally within the site's functional area (English skills required for this position)
- Participates in testing and evaluation of improvement suggestions
- Participates actively in development activities locally and globally (e.g., Kaizen, daily process huddles, etc.)
- Participates in the development and enforcement of functional content in training material together with the local job trainer

The **local job trainer** (LJT) takes the lead on the following tasks:

- Prepares, trains, motivates, and evaluates employees and vendors (using the defined training methods, material, and documentation)
- Follows up on trainees and ensures that they follow the local work standards with a focus on safety and quality
- Participates in networking activities locally within the site's functional area
- Develops and enforces the functional content in the training materials
- Collects improvement suggestions from the workforce and brings them to the global job trainer and FAMT for testing, evaluation, and approval
- Participates actively in development activities (e.g., Kaizen, daily process huddles, etc.) for self-development and development of the work area.

These were the initial three roles we created for operations, but while running the pilot project, we realized that we had missed one position when we saw how important it was to have someone take care of the writing tasks for the standards to have discipline within the documentation process. For many of our trainers, it was not a usual task to prepare written training material in addition to conducting the training itself. It is one thing to be able to do a job yourself, but it is something quite different to be able to train others in the same job. This requires that the trainers step out of their daily operational role and look into and analyze what they do and how they can communicate and teach it to others. So, we stressed the importance of having a technical writer role in the organization to alleviate the trainers from the struggle of working with the written documents both in their local languages and, especially, in English.

Here is the description of the role of the **technical writer**:

- Has responsibility for taking the technical language and making it easy to read and understand for the target user group
- Has responsibility for transforming practical knowledge and experience from line specialists into clear documentation
- Ensures alignment in documentation in relation to terminology, detail level, and so on
- Updates documentation and formats on SharePoint and at the workplace
- Supports the FAMT in practical work (e.g., administration, communication, etc.)
- Ensures consistency of translations to/from English or local language with the technical terminology of the LEGO Group and current updates of the technical terminology database

All the roles described so far are established and positioned in the functional areas, for example, in molding. In contrast, there is the **Learning Center consultant** (an internal LEGO "consultant"), who is located in the HR area and plays a separate role building up the training organization. The main areas of responsibility for this role are as follows:

- Supports in selecting, building, and developing the job training organization
- Develops, executes, and evaluates the Learning Center tools and processes
- Optimizes training and learning methodologies and documents and maintains templates, including global sharing
- Drives networking activities locally and globally

Another important role in the training organization is the LEGO Continuous Improvement (LCI) internal consultant. Our Lean specialists were not originally directly involved in the pilot process. However, they play an important role supporting the FAMTs with training on how to do daily huddles, the global and local job trainers with process confirmation, and the technical writers with setting up and maintaining standard documents.

In the establishment of the training organization, the critical part of the construction was to replicate the training organization at all LEGO production sites. This means that we have the same roles, the same job

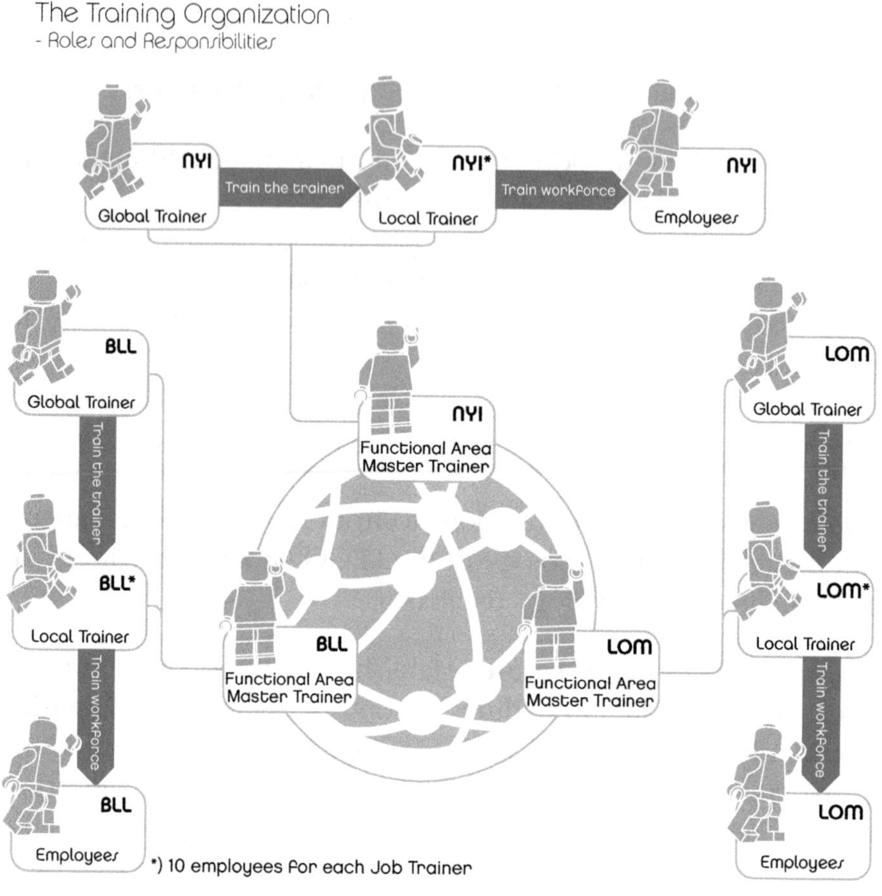

Figure 6.3 Global training organization.

descriptions/position profiles, and the same objectives and KPIs for each specific role. For example, when a global job trainer in molding from Hungary is collaborating and sharing knowledge with a global job trainer in molding from Mexico, they have the same job descriptions, they have been through the same training, and they are working according to the same standards. In Figure 6.3, you can see a depiction of the training organization in three molding locations—Denmark (BLL), Hungary (NVI), and Mexico (LOM)—which secures a strong foundation for the collaboration and standardization of the job training program. This is key to establishing and locking in, which we like to call "anchoring," the knowledge and skills at each site.

As noted previously, a standard or process used to be trained by a specialist from Denmark and pushed out to a new production site. With this new training organization, we have equal roles and share knowledge across sites, making training improvements in close global collaboration with each other at each site. This means that Hungarian, Danish, and Mexican trainers are equally involved, and the training is taking place in their local languages, avoiding misunderstandings. This way of replicating and developing the training organization secures equally high levels of commitment, dedication, and ownership from all of the local production sites. Furthermore, by creating a formal network, the trainers always know who to reach out to in case they are in doubt about standards, new processes, and so on. For example, within the packing area, a network of master trainers/global trainers always knows who to contact with questions, and this has been helpful and supportive for the production sites.

Who Wants to Be a Trainer?

A big part of the success of this new training organization would be getting the right people to apply for and hold the different trainer roles. For that, it had to be attractive and seen as a special position to be a LEGO job trainer. So, we needed to make visible and available the benefits of being a trainer as well as the possible career aspirations people might envision by holding the position. Many efforts were made to visualize the development program that candidates would follow to become a trainer, both to make new trainers feel confident they had made a good choice and to motivate and attract more people with the right talents.

As you can see in the career path model in Figure 6.4, it is possible to advance from being an operator to becoming a local job trainer and so on both within the same function and cross functionally. We have examples

Figure 6.4 Trainer career path model.

Figure 6.5 New trainer group in Hungary plant.

of global job trainers aiming at and becoming FAMTs and situations with some of our local trainers becoming frontline leaders. In some areas, trainers have to decide and choose to be a trainer at the expense of perhaps gaining another role, such as a safety coordinator, technology specialist, or some other similar position in the organization, so it is crucial to make the trainer's role attractive.

An example of how we try to make being a trainer an attractive position is that we do not allow just anyone to take the 10-hour Job Instruction (JI) class; it is not an open course and is not available in the LEGO training course catalog. We conduct the TWI training only for selected trainers, and we publicize the new trainers, showing them on the TV screens in the production area or putting their pictures on bulletin boards. Figure 6.5 shows a picture of a training team from the Hungary plant's molding areas.

Building Up the Training Organization: Selection Process

This section elaborates on the trainer development process. We are building up our training organization based on four main processes that are supported and guided by the Learning Center consultants: selection, onboarding, development, and evaluation (see Table 6.2). As we consider the selection and development processes to be the two most vital parts of this effort, we provide greater detail about these two areas and touch lightly on the other two.

The first step is the selection process. We consider this to be one of the most important activities in building up the training organization based on what we learned prior to the pilot project, where the best specialists were

Table 6.2 Four Main Processes for Building Up Trainers

Selection	Onboarding	Development	Evaluation
- Final scoping with mgmt. - Information to workforce - Application period - Interviews/exercises - Selection decision with mgmt. - Feedback to applicants	- Kickoff session (activities and roles) - Follow-up on functional expertise	- JI 10-hour class - Practice JI method and Job Breakdowns - Training & Learning Skills Course (adjusted to specific role)	- Evaluation on JI before training workforce - Quarterly evaluation on JI and trainer skills

not necessarily the best trainers. Furthermore, there was a strong request from the production director of the Mexico plant to intensify our efforts in this area and set up a specific process for selecting trainers that would be able to recognize the needed trainer talents. In this way, we could make sure that we were building on the right trainer profiles.

The initial activity is to do the **scoping** of specific training activities with managers in each area to determine what needs we were trying to address in the area. The needs can be based on things like growth and expansion, transfer of knowledge and processes, or implementation of new technology/equipment. The scoping also includes a common plan for the number of trainers, the timing of their development, and so on. Thereafter, we provide the department with **information** about the trainer role and how employees are to apply for the position. This is communicated through visual screens in the production areas and a half-hour "road-show" presentation we make to tell people about the job training program to inspire them to participate. The road-show approach works well. We usually put it on at the end of the daily huddle or other stand-up meetings where employees in specific functional areas gather. We also put up posters and explain about the program and how to become a trainer. Often, an existing trainer participates and answers questions from the audience.

Interested **candidates apply** within a specific time frame simply by writing an e-mail with a couple of lines explaining why they are interested in becoming a trainer. If candidates with high potential are not applying by themselves, leaders in their areas can give some subtle hints to encourage these individuals to think about the opportunity. Then, the next step is the **interview and exercises**; we analyze the talents of each candidate and look for their potential as a trainer. These meetings are conducted by the Learning Center with the participation of the FAMT of the area. In setting up the selection process, the managers of the area are assessing the trainer candidate's job-specific knowledge and skills. In addition, we make an assessment of their trainer competencies (knowledge, skills, and behavior) based on criteria described in *Toyota Talent* (Liker and Meier, 2007, pp. 63–71).

How to Spot Trainer Talents in the Organization

Assessing the potential of the candidates to be a trainer is the most challenging part of the selection process. To make the best choices, evaluators must know what they are looking for in a good trainer from numerous angles, including both technical skills and abilities and personal characteristics. We distinguish between natural trainer talents, like patience and genuine caring for others, and fundamental learnable skills, like observational skills and effective communication. In other words, we can teach someone better methods of communication, but, while not impossible, it is tough to "teach" someone to be more caring of others. While we can try to develop these personal characteristics in the trainer, it is easier if they are a natural part of the talent and personality that the person brings to the position. Fundamental skills, on the other hand, are viewed as learnable skills and abilities that can be gained through continuous practice and guidance. We can list these trainer competencies in these two categories:

Fundamental Trainer Skills and Abilities:

- **Observation and Analysis Ability:** Has ability to assess current work methods, trainees, results of training, and so on
- **Communication Skills:** Has clear/concise speech, gestures, body language, active listening, and the like
- **Attention to Detail:** Provides thorough/complete preparation prior to training, critical details of the job, and so on
- **Job Knowledge:** Has solid knowledge of the job, quality issues, tricks/key points/critical issues about the job
- **Respect for Fellow Employees/Colleagues:** Demonstrates respect for others; act as a hard-working, knowledgeable role model; shows genuine caring for others

Intuitive Ability/Personal Characteristics of the Trainer—Natural Talents Within:

- **Willingness to Learn:** Has desire and ability to continually learn and grow and to reinvest one's own learning into teaching
- **Adaptability and Flexibility:** Is able to adapt to work situations, trainees, conditions, and so on

Care and Concern for Others: Genuinely desires trainees to become successful in the job and as a person; offers help naturally without being asked

Patience: Is able to cope with different trainee abilities

Persistence: Can stick to learning process until desired output is achieved and understands different abilities/needs of trainees

Responsibility: Is able to take many decisions throughout the learning process

Confidence and Leadership: Is confident (e.g., when challenged, leads the learning process and provides guidance)

Questioning Nature: Questions the content of the job and understands each part of it; is able to understand/answer questions from trainees

These descriptions have worked well for us in explaining to the trainers what we expect from them; we have used them to clarify and discuss trainer skills and personalities. This is the way we communicate our expectations to the trainers. However, to reduce complexity in the selection process, we have merged the skills and natural abilities into six main categories: work attitude, confidence and leadership, communication skills, social attitude, cognitive skills, and questioning nature (see Figure 6.6).

The aim of the selection process is to identify the competencies that can be evaluated in the situational/practical exercises and interview we do with the candidates and translate them into observable behavioral signs or manifestations—to know what we are looking for when we observe them in a practical exercise. The following guidelines were developed by our Learning Center consultants to recognize these competencies when selecting the best candidates to become trainers. For each of the

Figure 6.6 Categories for spotting trainer talent.

six main skills and abilities categories, we spelled out what an average person would look like under that criterion and what behavior a below-average or excellent person would exhibit under the stated conditions. Following these guidelines, the Learning Center consultants, in the role as interviewers, would then be looking specifically for these things in each candidate, and they could evaluate each candidate effectively and fairly on each competency.

Cognitive Skills

Observing and Analytical Skills; Attention to Detail

Definition: The extent to which one can think in a structured and systematic way. The way one collects information, organizes the pieces into a coherent and meaningful system, and then presents it in a structured, understandable way.				
Description of Scale and Grades in Terms of Behavioral Characteristics				
Not Sufficient	*Below Average*	*Average*	*Above Average*	*Excellent*
Not able to understand the operation of the work procedure and functions of certain Important Steps and Key Points and the importance of Reasons within the procedure.		More or less able to understand the operation of the work procedure and some of the functions of certain Important Steps but little of the importance of Reasons for Key Points within the procedure.		Completely understands the operation of the work procedure and the functions of certain Important Steps and Key Points and comprehends the importance of Reasons for Key Points within the procedure.
Does not pay appropriate attention to obtain and present all the available information in a precise manner.		Amount of attention he or she pays to obtain and present all the available information in a precise manner varies.		Pays appropriate attention to obtain and present all available information in a precise manner.

Continued

Observing and Analytical Skills; Attention to Detail

Confuses, oversimplifies, or even misses important details within the process (Important Steps, Key Points, and Reasons); hence, he or she is not successful in transferring knowledge, and it is not easy to follow the candidate.		The Important Steps, Key Points, and Reasons within the process are mentioned and presented in an understandable way and in the correct order.		All the Important Steps, Key Points, and Reasons within the process are mentioned and presented in a clearly understandable way and in the fully correct order.
Can be confused by clarifying questions; not confident in the correctness of information he or she has gained.		Sometimes can be confused by clarifying questions, but overall is confident in the correctness of information he or she has gained.		Cannot be confused by clarifying questions; is totally confident in the correctness of information he or she has gained.

Communication Skills

Verbal Communication

Definition: The way one can convey or transmit information to another person with the help of verbal tools.				
Description of Scale and Grades in Terms of Behavioral Characteristics				
Not Sufficient	*Below Average*	*Average*	*Above Average*	*Excellent*
Does not communicate well; not able to convey his or her thoughts properly; speaking is without any meaningful content.		Can communicate at an average level; usually is able to tell his or her thoughts and ask questions in an understandable and clear way.		Is an excellent communicator; can express his or her thoughts in a completely clear and understandable manner.

Continued

Verbal Communication

His or her way of telling information is confusing, uncertain, and hardly understandable and is full of mistakes regarding formal traits.		Uses vocabulary of an average quality but does not possess and apply extra tools to make his or her communication more efficient.		Uses a large variety of verbal and nonverbal tools (repetition, colorful vocabulary, regulation of intonation, emphasis and volume, eye contact, intense gestures, etc.) to make his or her communication efficient; is able to adjust personal style to the characteristics of the partner.
Suggested subcategories within verbal communication to observe: • Structure (clarity and understandability of communication) • Content of communication (richness, accuracy, and concreteness of information) • Formal signs of communication (vocabulary, articulation and pronunciation, grammatical formulation)				

Nonverbal Communication

Definition: The way one communicates attitudes and affective states (emotions and feelings) toward other people with the help of nonverbal tools, such as body language (mimics, gestures), voice modulation (intonation, volume, stressing, rhythm), and proximity (regulating social space).				
Description of Scale and Grades in Terms of Behavioral Characteristics				
Not Sufficient	*Below Average*	*Average*	*Above Average*	*Excellent*
The nonverbal behavioral signs shown are rarely in line with the content of his or her verbal communication.		The nonverbal behavioral signs shown are usually in line with the content of his or her verbal communication.		The nonverbal behavioral signs shown are always in line with the content of his or her verbal communication.

Continued

Nonverbal Communication

Communication about his or her affective states does not seem to be aligned with the partner and the context of the situation. Therefore, the candidate does not really evoke the impression of a credible person but rather appears dishonest or misleading.		Communication about his or her affective states sometimes seems to be well balanced and aligned with the situational factors, but at other times there is a discrepancy between his or her verbal and nonverbal communication.		Communication about his or her affective states seems to be well balanced and aligned with the partner and with the context of the situation. He or she evokes the impression of a credible person.
Suggested subcategories within nonverbal communication to observe: • Adjustment to partner in content and style (connect thoughts to the words of the partner, align his or her communicational style with the other person's personality and behavior) • Ability to attract attention (make the partner be interested in the topic through the style of speaking) • Social-emotional communication (the way of approaching others, gestures, balance in emotional state, eye contact, etc.)				

Social Attitude

Respect for Others (Including Cooperativeness and Flexibility)

Definition: The direction (positive/negative) of the attitude with which one usually relates toward other people and the extent to which one acts in social situations in a way that shows attention, care, and concern about others' interest and well-being. This often manifests itself in the form of cooperation and flexibility in adapting to the given circumstances developed by others' demands and needs.

Continued

Respect for Others (Including Cooperativeness and Flexibility)

Description of Scale and Grades in Terms of Behavioral Characteristics				
Not Sufficient	*Below Average*	*Average*	*Above Average*	*Excellent*
His or her attitude towards others is usually rejecting, offending, or even aggressive. Does not show a respectful manner involving interactions with others.		Shows no interest in forming and maintaining a balanced and psychologically comforting relation with others.		Shows respect for other people involving interaction with them.
Not concerned about how to help others; often rejects requests, and sometimes loses his or her temper because of lack of patience.		Does not express much willingness to help, however, if asked, he or she serves others' requests.		Expresses genuine care about the well-being of others. Demonstrates high willingness to help, even if not asked to do so.
He or she is not curious or interested in the partner's opinion at all.		Shows some interest in the partner's views and opinions.		Tries to understand the partner's opinion with empathy and strives to consider his or her requests.
Usually rejects and even degrades the appropriateness of the partner's thoughts. Obviously does not intend to consider and understand the other's viewpoint.		Expresses his or her understanding, but this does not reflect true empathy.		Turns to others with attention and listens to their views carefully. His or her general attitude toward people is characterized by kindness, patience, acceptance, tolerance and helpfulness.

Continued

Respect for Others (Including Cooperativeness and Flexibility)

Cares only for his or her own interest and suggests solutions that are contrary to the other's needs.		Makes few efforts to find an outcome which respects all the participants' purposes and interests, but is able to accept a proposal for it.		Often searches for solutions which are beneficial for all participants in the situation.

Confidence and Leadership

Confidence and Leadership (Also Including Influence on Others)

Definition: The degree to which one is able to put his or her intentions across and convince others of certain things throughout a diverse scale of social situations. When realizing personal goals, one can rely on this skill to step up confidently in front of a bigger audience, be willing to take leadership over a group of people, and be willing and able to have an influence on others.				
Description of Scale and Grades in Terms of Behavioral Characteristics				
Not Sufficient	*Below Average*	*Average*	*Above Average*	*Excellent*
Is not able to have an effect on others' views through persuasion; fails to convince other people of a certain idea or concept.		How successful he or she is in convincing others of the appropriateness of his or her ideas varies; candidate uses some tools for the sake of persuasion, but not a big variety of them, and does not endure for long, especially if perceives resistance.		Is able to convince others of the appropriateness of his or her ideas with the help of good skills of persuasion (explains persistently, approaches topic from diverse perspectives, refers to reliable sources and to own experiences, etc.).

Continued

Confidence and Leadership (Also Including Influence on Others)

Most of the time his or her actions and behaviors demonstrate uncertainty, ambiguity, and lack of self-confidence.		Shows a medium level of confidence and determination when needs to speak out for his or her interests; sometimes becomes uncertain about telling his or her opinion.		Has the ability to stand out with a proper amount of confidence and assertiveness and give opinions with a high level of determination.
Is not motivated to invest energy into asserting his or her interests; usually gives up on intention and will when faced with difficulties from the partner's side.		Hesitates to assert his or her interests, especially when perceives resistance. But takes control when in a good position to do so.		Has the motivation and energy to put his or her interests across, even if he or she perceives resistance from the partner's side.
Shows no motivation and ability to take the lead over another person or group of people; prefers to be rather passive and to have the position of a subordinate.		Demonstrates relatively big amount of changeability in the motivation and ability to take the lead over another person or a group; sometimes is prone to provide guidelines and give instructions to others, but at other times lets control out of his or her own hands.		Is prone to take the chance to lead another person (or a certain group of people) toward a goal that is commonly beneficial for both of them.

Work Attitude

Performance-Motivation (Including Persistence and Responsibility)

Definition: The degree to which a person is motivated to contribute persistently to a final outcome with one's own efforts, to which one is facilitated and motivated by one's own results, and to which one takes responsibility for one's acts.

Description of Scale and Grades in Terms of Behavioral Characteristics

Not Sufficient	*Below Average*	*Average*	*Above Average*	*Excellent*
Does not invest energy and makes only minor efforts to perform at an acceptable level.		Invests an average amount of energy into his or her work that is enough to achieve a medium-level performance.		Invests a lot of energy into his or her work to achieve a high-level performance.
Is not persistent in reaching his or her aims when faced with difficulties.		To a certain extent, the candidate tries to endure and cope with difficulties but after a while tends to give up and choose an easier way.		Is persistent in reaching his or her aims even if faced with obstacles; does not give up his or her targets when coping with difficulties; works diligently on personal goals.
Does not set motivating goals for himself or herself; usually defines deliverables under the level of his or her competencies, which indicates a low aspiration level.		Sets targets that are realistic in respect to his or her capabilities, but these goals do not motivate the candidate to improve performance.		Is able to set realistic and challenging goals to ensure appropriate level of accomplishment.

Continued

Performance-Motivation (Including Persistence and Responsibility)

Does not feel ownership for the solution of the situation and rejects taking responsibility for contributing to the outcome.		Satisfied with the solution if it puts him or her into a prosperous light, but takes responsibility only for own beneficial and advantageous acts; does not take ownership for possible mistakes.		Is proud of and motivated further by the performance level he or she achieved and takes responsibility for his or her acts leading to the outcome.

Questioning Nature

Questioning Nature and Initiative

Definition: The degree to which one invests energy and shows activity throughout the completion of a task or the solution of a problem. This includes the motivation to challenge the given situation and others' views by asking questions that can encourage shifts in perspectives.

Description of Scale and Grades in Terms of Behavioral Characteristics

Not Sufficient	*Below Average*	*Average*	*Above Average*	*Excellent*
Does not play an active role in the situation; is passive most of the time; does not share his or her thoughts and ideas with others and asks questions rarely.		His or her level of initiative shows variability; sometimes is willing to share thoughts, other times only does so when asked to give an opinion.		Shows a constant level of initiative by continuously asking clarifying questions, sharing ideas, and telling opinions and arguments in favor of them.
It is not important for him or her to participate in situations and contribute actively toward the outcomes and solutions by investing personal energy and capability.		The intensity of his or her participation is changeable; takes part actively once, then acts more reserved.		Is eager to take part actively, with his or her own personality and knowledge of the situation.

Continued

Questioning Nature and Initiative

Rarely fights for his or her viewpoint and does not question others' proposals.		Usually reasons and argues in favor of his or her viewpoint, but questions others' proposals less frequently.		Not afraid of challenging others' views to find the optimal outcome and avoid possible errors.
His or her proposals and ideas are vague and oftentimes off topic.		His or her proposals and ideas are rather schematic, conventional, and general.		His or her thoughts and ideas are usually relevant, creative, and authentic.
In case of a change in the situation, he or she reacts slowly, with few relevant comments.		Reaction is variable when a situation changes; sluggish to react but participates in giving comments.		In case of a sudden change in the situation, he or she reacts promptly and skillfully.

The Interview

The competencies described are the foundation of the selection process. They are used as guidelines by the interviewers during the interview and exercise part of the selection process so that a good assessment can be made on the selection of the most appropriate people to become trainers in the LEGO Group. In any specific selection session, we go through three steps, an interview and two practical exercises. Table 6.3 shows an overview of the process.

In the selection session, the first thing to be done is to set the stage and give the candidates confidence by explaining that the process is an informal session where we encourage their questions. Then, we introduce the candidates to what is going to happen in the session. The first part of the session is the interview. We set up the interview to consist of a combination of open-ended questions the prospective trainers can reflect on and give their own answers as well as semi-open-ended questions for which guidance is provided to help them give more specific answers.

These questions make it possible to explore essential trainer competencies, behavior, and attitude to determine the strength of the candidates and areas of needed development. The questions are situational, meaning that

Table 6.3 Selection Process: Interview and Practical Exercises

Interview • Situational questions about job training (adjusted to candidate's experience); for example, What has been the most difficult training situation that you have had to deal with? How did you handle it? **Focus on behaviors/skills/attitudes**
Practical Exercise 1 • Analyzing and presenting text (procedure). • Arrange text in proper/logical order and present/introduce to "trainee." **Focus on observing/analytical skills/communication**
Practical Exercise 2 • Building small LEGO model; instruct relatively simple job without seeing what the trainee is building. **Focus on instruction skills, confidentiality in trainer role, adjustment of trainer style with the "trainee," and so on.**

the trainers must image specific trainer-related situations to properly answer. We also make the questions relevant to their functional areas and prepare specific examples in collaboration with the area master trainer to ensure that the candidates can relate to the questions asked. Here are some examples of questions we ask:

- Imagine a situation where you are going to train a colleague from Mexico in doing the job of measuring a mold part. Try to explain how you will prepare this training session and what you will do before the training.

This question is related to the "Attention to Detail" (thorough/complete preparation prior to training, critical details of the job, etc.) parameter of the Fundamental Trainer Skills and Abilities category of trainer competencies.

- Trainers should be role models. Can you explain how you as a trainer can be a role model and a good example for your colleagues? Try to give some examples.

This question is related to the trainer competency category of Intuitive Ability/Personal Characteristics of the Trainer and the parameter "Respect for Fellow Employees/Colleagues" (demonstrates respect for others; acts as a hardworking, knowledgeable role model; shows genuine caring for others).

- Imagine you are in a training situation where one new employee is involved in a 3-day training cycle. After 2 days of training, you realize that you have only reached one-third of the training program. Try to explain how you would handle the situation.

This question is related to the parameters "Adaptability and Flexibility" (is able to adapt to work situations, trainees, conditions, etc.) and "Taking Responsibility" (is able to take many decisions throughout the learning process).

- Try to imagine a difficult training situation that you had to deal with. What did you do to overcome it?

This is an open-ended question. If it is too difficult and the candidate is struggling with finding a good example, we guide him or her by giving a more specific, but still relevant, example that can be formulated in a more concrete way:

- You are in a training situation with a small group of experienced mold setters, and you are training them on a new changeover process that is to be implemented. The experienced setters challenge the new process and argue that they prefer the old way. They claim that the new process will not work. Try to explain how you would handle the situation.

These questions are related to the parameter "Confidence and Leadership" (is confident; e.g., when challenged, leads the learning process and provides guidance).

Training employees can present many challenges, and some common difficulties include lack of manager or supervisor support, lack of interest or motivation from the employees, and maintaining self-motivation when progress is slow or faltering. Knowing these challenges will be confronted if the candidate is selected to be a trainer, we try to find out how they will handle them.

- What skills and abilities do you bring to the trainer position: What do you bring with you to this job?
- Try to explain your reasons for wanting to become a trainer: Why are you interested in training others?

These questions are related to the parameter "Willingness to Learn" (has desire and ability to continually learn and grow and to reinvest one's own learning into teaching).

As you can see from the nature of all these questions, throughout the interview we refer to the trainer skills and competencies: job knowledge, communication, adaptability, patience, and so on. Or, to put it another way, the interview questions were designed around these parameters to give us the feedback we needed to make a good evaluation. Furthermore, in relation to the answers to the questions about their reasons and interest in becoming a trainer, it is interesting to observe whether their main focus is on their own development process and the opportunity to learn new skills and obtain more knowledge as a trainer or whether they are focused on sharing their knowledge with others and the learning process of others. Of course, what we are looking for is a good balance of both of these motivations.

Practical Exercise 1

Following the interview, a practical exercise is introduced. The trainer candidate has to play the role of a job trainer receiving a document full of information about how to complete a certain job (work procedure). The sheet contains all the data necessary for task solving, but not in the correct order. The candidate's assignment is to read the text and arrange the pieces of information in the proper sequence so that they make up a logical description of the job. The next step is to present the work procedure. The exercise is introduced in the following way:

> **Instruction**
>
> Please, imagine yourself in the role of a job trainer who has been working in the molding area for 3 years. As a trainer, your basic tasks include teaching newcomers both theoretical knowledge and technical skills, introducing them to the work processes, and getting them to practice and gain experience in the work tasks. Besides this, you are responsible for helping them become acquainted with and accustomed to their new work environment by introducing them to their new colleagues, guiding them around the most important places within the plant, and informing them about the company rules, roles, and customs accepted and followed by every coworker.
>
> Imagine that a new colleague has just arrived to your area, and the two of you are setting up to go on the regular "factory tour," during which you

> will show him where he can find and how he can use the changing room, the cafeteria, the restrooms, and the kitchenette. Arriving in the kitchenette, your most important task is to explain to the newcomer how to make coffee. This is essential because at your company a big emphasis is put on completing all kinds of activities in a way that is most efficient and that ensures the workers' safety. This also goes for coffee making. You have already prepared to describe this process in detail, but unfortunately the computer on which you stored the relevant document broke down at the last moment, and you had to borrow the overview used by one of your colleagues. This file, although it contains all the necessary information, is mixed up and confusing because the data involved are not presented in the correct order. Hence, you have to create the proper sequence for the pieces of information to tell the process of making coffee precisely to the novice.
>
> You have 15 minutes to prepare your presentation based on this written procedure (see Figure 6.7). You have the possibility to make notes during the preparation and then afterward use these notes throughout your presentation. After the 15 minutes have passed, you will be asked to present the job to the new colleague in 10 minutes.

With the help of this exercise, we can see how logically the candidate can organize information into structures on a mental level, how skilled he or she is in presenting this structure of data to the audience in a clear and understandable way, and how much energy the candidate is willing to invest in solving a challenging task.

How to Make a Cup of Coffee

We should not pour in too much milk; otherwise, the coffee dilutes and loses its special aroma.
Proper sanitation is important to avoid the spread of bacteria, hence reducing the risk of illnesses.
When we are ready with the coffee, place the mug on a tray.
Put instant coffee and sugar in the mug.
To determine the appropriate amount of water to add for our taste, add small amounts of water two or three times and taste after each addition.
Also, hold the mug firmly, without shaking to prevent spilling the coffee.
We can also pour milk into our coffee.
Use only a clean mug and spoon.
Use hot water for the coffee.
Make sure the milk is fresh, it helps us avoid illnesses.
Hold the cup by its handle because the mug is hot and can burn our fingers.
We should add 1 teaspoon each of both the sugar and the instant coffee because generally it is the amount that provides the most delicious flavor.
Cap milk before returning it to the fridge. Use a cup and spoon from the cupboard and not from the counter, to ensure it is clean.
Slowly pour hot water in the mug to prevent splashing of hot water and burning your hand.
At last—enjoy your refreshment.

Figure 6.7 Instructions on how to make coffee.

The competencies we evaluate in this exercise are

- Observation and analytical skills; attention to detail
- Verbal communication (content, structure, formal signs)
- Nonverbal communication (adjusting to partner, attracting attention, social-emotional communication)
- Performance-motivation (persistence and responsibility)

Practical Exercise 2

In the second exercise, the candidate is asked to teach someone a relatively easy job that does not really require any previous knowledge or preparation. Here, it is about building a simple LEGO model out of a few bricks with the help of a manual (instruction sheet). Other similar simple tasks could also be used.

Instruction

In this exercise, you will have the task of teaching a relatively easy job to another person; namely, you are going to train someone how to build a simple LEGO model out of a promotion bag with bricks. You are going to be provided with the needed bricks and a short manual instructing how to build the target model. However, there is one condition: You cannot see what your trainee is doing. The goal is for your trainee to build the model based solely on your verbal instructions. You are allowed to use any tools you think are useful for solving the task as long as you make sure you cannot see anything your trainee is doing.

With the help of this exercise, we can see how clearly the candidate is able to instruct another person, what kind of instructor skills the candidate possesses, how confident and persuasive he or she is in the role of a trainer, how well the candidate can adjust his or her style to the trainee's personality, how adequately and empathetically the candidate can react to the trainee's needs and behavior, and how much energy he or she invests into reaching as high a level of performance as possible.

The competencies we evaluate in this exercise are

- Verbal communication (content, structure, formal signs of communication)
- Nonverbal communication (adjusting to partner, attracting attention, social-emotional communication)

- Respect for others and cooperation
- Confidence and leadership
- Performance-motivation (persistence and responsibility)
- Initiative

After this last exercise, the trainer candidate is informed of the next steps in the process and the timing of the forthcoming feedback. Again, questions are encouraged.

During the entire selection session, notes are made to create a summary of the assessment for each individual candidate. Furthermore, a sheet is filled out based on each exercise, and specific examples, quotations, and so on are documented for use during the feedback session with the candidate and the later dialogue with the leader of the area. The summary report sheet is shown in Figure 6.8.

After the selection process is complete, we have a meeting with the managers where we share our findings from the interview and exercises. Based on the selection process results, the Learning Center consultant recommends who they feel the future trainers should be, but it is the manager who makes the final decision for his or her department. Finally, all the candidates are given individual feedback, in some cases together with the manager, on how the process went and the final decision.

Onboarding Process

Now that the selection process is complete, we move on to the onboarding process and the **kickoff** (refer to Table 6.2). The purpose of this part is to make sure that the newly selected trainers understand exactly what their new roles will be and the process of development activities they will be going through to become full-fledged trainers. For many trainers, this will be a brand new experience, and some will feel insecure about their new role, so this gives them an overview of their future activities and what is expected from them to perform the job. Moving forward to the development process, we will ensure that the new trainers have a sufficient level of **functional expertise**, and that they go through any relevant training and practice periods to ensure they are fully up to speed with the techinical aspects of the job before learning how to become trainers.

LOCAL TRAINER SELECTION PROCESS

CANDIDATE'S GLOBAL REPORT

CANDIDATE'S NAME	
EVALUATOR'S NAME	
DATE	

COMPETENCY	SCALE				
	Not Sufficient	**Below Average**	**Average**	**Above Average**	**Excellent**
Cognitive abilities					
Observation & Analysis					
Communication skills					
Content of communication					
Structure of communication					
Formal signs of communication					
Nonverbal communication					
Proximity					
Attract attention					
Social-emotional communication					
Social attitude					
Respect for others					
Act as a role model					
Adaptability & Flexibility					

Figure 6.8 Summary report on trainer interview and selection.

Confidence & Leadership					
Questioning Nature & Initiative					
Work Attitude					
Willingness to learn					
Identification of Training needs					
Performance-motivation					

CONCLUSIONS (evaluators only)

STRENGTHS	AREAS TO IMPROVE
*	*
*	*
*	*
*	*
*	*

- Would the candidate fit the position profile?

- Would the candidate develop competencies through training program?

- Which topics could you conclude as necessary for the training program of local trainers?

Evaluator´s name	Local Learning Center Consultant

Figure 6.8 (*Continued*)

Development Process

The **development process** is targeted toward the different roles: local job trainers, global job trainers, FAMTs, and HR Learning Center consultants. In the model shown in Figure 6.9, you can find the different development activities for each of these roles.

All the training and development activities are set up in accordance with the LEGO learning principle: 70-20-10. This means that 70% of your learning shall

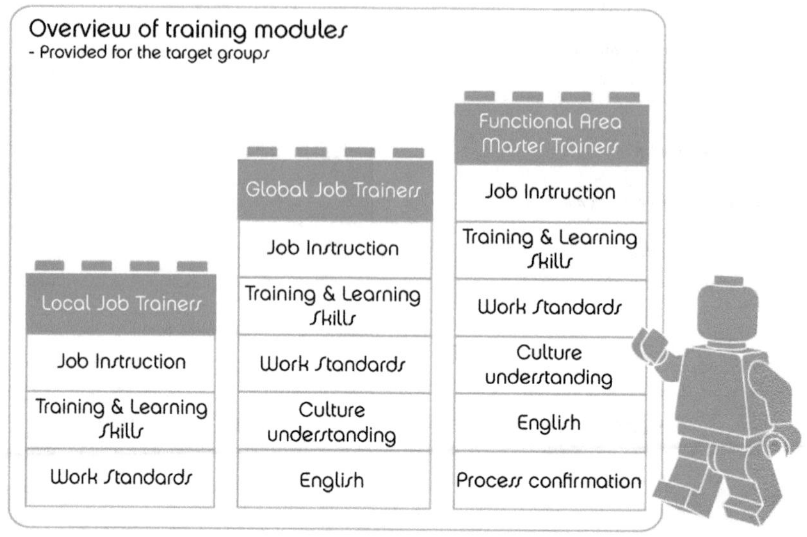

Figure 6.9 Development process for each trainer role.

take place through development on the job, 20% by coaching and mentoring and daily contact with colleagues and managers or other specialists, and the last 10% of your learning and development activities will be formal learning such as classroom education and other courses. A good example of this principle is the JI skill training, for which the 10% of the formal classroom training is the 10-hour class, which is followed by support from the Learning Center and FAMTs. Here, we add a lot of practice until the person masters the method.

Job Instruction

We start by giving the TWI JI 10-hour class to the trainers, which is a vital cornerstone of the trainer development program. The 10-hour class is conducted by Learning Center consultants who have completed the 40-hour Job Instruction Trainer Development Course certification program given by the TWI Institute. After the certification, which is conducted in English, the Learning Center consultants are provided translations of the JI training materials in their local languages so that they can deliver the JI training in the local language: Hungarian, Czech, Spanish, and Danish.

The JI training is followed by much practice, with the focus on learning to make Job Breakdowns while practicing the method. Before we send these new trainers out to use the JI skills they have learned, we must ensure that they have mastered the method. Here, we learned that it works well in the beginning to have some completed simple Job Breakdowns to practice on so they can focus only on following the instruction method. After they master the four-step method of giving good instruction, then they can progress to the more difficult skill of breaking down a new job for instruction. But, all in all, we maintain a synergy of both developing the Job Breakdowns and practicing the method. Here, FAMTs, global trainers, technical writers, and LCI are all supporting the process.

It quickly became clear, as discussed in Chapter 4, that the hard part to learn in JI was good Job Breakdown skills. Some people pick up the skill easily. But for most, it takes time and experience to acquire. There was one example from the verification area, working with measuring, where one of our trainers picked up the breakdown skill quickly after being carefully coached. Now, she is setting the direction for the quality of the Job Breakdowns on the team. It is beneficial, in this way, to have some Job Breakdown "champions" in the areas to take the lead on setting a high standard for this process. Figure 6.10 shows an example of Job Breakdown from the verification area working on measuring.

Originally, in the pilot project, the 10-hour class was only given to trainers who would actually be instructing the jobs. But, we found that it is important to provide the JI training to leaders and supervisors in the area as well so that they gain an understanding of the method and are able to fully support the instruction effort once it begins. Having this insight into the training method and the ability to have a dialogue with the trainers about the method, the approach, and the expected impact from the training proved essential in reaching the final results.

Training and Learning Skills Course

Another important part of the trainer development process is the Training and Learning Skills Course. We typically conduct this training 1–3 months after the trainers have received the JI class so they have time to practice the method and gain confidence in the JI methodology prior to learning something new. We also make sure that the trainers have been through this course before they conduct their first training. The program consists of four modules that provide insights into adult learning and into their own behavioral patterns, communication, and the like. The program has been developed in global collaboration with the various worldwide LEGO organizations and their respective Learning Center staff.

Prior to the pilot project, it was discussed whether external consultants needed to be brought in to support setting up this part of the program.

LEGO	Dial gauge Job break down sheet	
Prepared by: **Benthe Felthaus**	**Document owner:** **Per R. Schmidt**	**Date of implementation:** 15/03-2012 **Date of change:** 07/06 - 2012 **Estimated training time:** 60 min.
Parts:	Mould part and drawing	
Tools & Materials:	Dial gauge, lint free-cloth, JFA clean, calibration wheel	

No.	Importants steps A logical part of a process - when something happens that promotes the job	Key point Anything in a process which: 1. Ensures or spoils the job 2. Hurts the employee 3. Makes the job easier to carry out, e.g. hints, tricks, special timing, any small special info	Reasons of key points The reason of key points
0	Common key point	Read calibration no. on the equipment Match the date and color on calibration wheel	To make sure that the measuring equipment has been calibrated
1	Prepare dial gauge	1. Add JFA clean 2. Wipe off plane with a lint-free cloth 3. Tightened the key 4. Arm/spindle has to run easily 5. Clean probe with clean lint-free cloth	1-2. To reach correct dimensions and to remove dirt and grease 3. To secure 4. Correct measuring 5. Correct measuring
2	Switch on/zero set	1. Turn on/off 2. Loosen finger screw 3. Move carefully on plane 4. Tighten finger screw 5. Press zero	1. Preparing the dial gauge 2. To move up and down 3. To avoid damge 4. To make a zero point 5. To set zero
3	Insert a mold part	1. Clean mold part with lint-free cloth 2. Lift arm/spindle and put in the mold part, don't touch where we measure 3. Carefully put down arm/spindle 4. Read dimensions	1. To remove dirt and ensure correct dimensions 2. In order to measure the dimension 3. To avoid damaging probe/spindle 4. Final dimension
4			

Figure 6.10 Sample of Job Breakdown Sheet from verification area.

But, from the first meeting of the various Learning Center staff to tackle this issue, it quickly became clear that with the great diversity in our backgrounds (psychologist, specialist in the didactical area, educational psychologist, and Lean specialist), there was no need for external advice or help. And, the great advantage of setting up the program "in house" was that it allowed the final product to be fully adjusted to the target group.

Leaning Center consultants from each production site first created drafts for the content, and these were discussed, adjusted, tried out, and finalized. We ended up with a complete program ready to roll out to all sites of the training organization, and because of their dedication to the effort, all of the Learning Center consultants feel a great commitment to and ownership of the program. In the model shown in Table 6.4, you can see an overview of the content, which consists of four modules.

Table 6.4 Training and Learning Skills Course: Modules

Module 1	Adult Learning and Learning Styles
Module 2	Introduction to DiSC Model
	Individual Feedback on DiSC Profile
Module 3	Communication and Feedback
Module 4	Training Challenges

Module 1: Adult Learning

The first module covers adult learning; here, the trainers are introduced to learning theories and adult learning patterns in a practical way through many exercises, examples, and dialogues. We are building on the knowledge and experience they already possess. We are also relating the theory to the JI method to make it relevant and to give the trainers the opportunity of taking a look at the JI method from a higher-level perspective.

To give an example, we introduce the social learning theory developed in the 1970s by Albert Bandura, which covers these four steps in the observational learning and modeling process:

1. *Attention*, including modeled events and observer characteristics
2. *Retention* (remembering what one observed), including, for example, symbolic coding, cognitive organization, and the like
3. *Motor reproduction* (ability to reproduce the behavior), including physical capabilities, self-observation of reproduction, accuracy of feedback
4. *Motivation* (good reason to want to adopt the behavior), including external and self-reinforcement

Based on these principles of social learning theory, we explain and discuss the link to the four-step method of JI as shown in Figure 6.11.

Attention: You learn by being attentive to the actions and behaviors that the role model is demonstrating. It is possible to select and pick out critical elements—being repeated many times—when you observe the role model. Attention can be caught when the observed behavior is highly effective (quick, smooth, without errors); is applied uniformly by many people over a long period of time; and is done in a positive way to motivate the observer to do the right things.

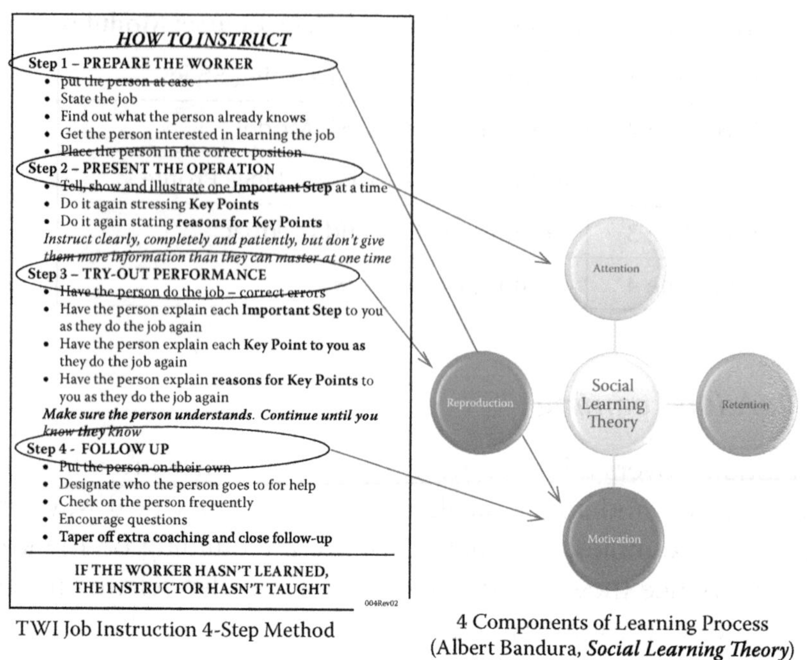

TWI Job Instruction 4-Step Method

4 Components of Learning Process (Albert Bandura, *Social Learning Theory*)

Figure 6.11 Job Instruction link to Social Learning Theory.

Retention: What is observed must be built on language and visual tools, which are then stored in your memory. It is important to focus on what is most relevant and when symbols and other schematic tools should be applied. The information is stored in the long-term memory when one consistent picture of the execution is made (step 2 of the JI method). Doing something is different from simply talking about the same thing; these two activities use separate parts of our nervous system. If we combine the different coding systems (verbal and psychomotor), they will enhance each other, and our knowledge or skill about the activity will be more conscious, complete, and confident.

Motivation: Motivation and attention are closely linked. Motivation is not given as a precondition but will be generated during the practice. Furthermore, there should be a good reason to do the task, which is often related to obtaining a good result (steps 1 and 4 of the JI method). Sources of motivation can be the desire or drive to be accepted by a group of people, to ensure safety by keeping the rules, to reach great performance through continuous improvement, to obtain the ownership and responsibility for a certain job, and more.

Reproduction: Now, the learner will try to integrate the symbols and schemes that are stored in the brain and turn them into new actions and behaviors that can be done on his or her own (step 3 of the JI method). This information will be taken from the long-term memory when the learner tries to do the task. With only a short time observing the role model, there is a risk of insufficient representation, which could lead to a different reproduction of the task. Therefore, it is important that the role model do the task several times, including, if needed, after the execution has started.

In this way, by comparing the theory of how adults learn with the practical process of instruction using the JI four-step method, we are creating an awareness of the *trainer's* execution of the training and importance of their approach to doing this effectively. This learning helps the trainers to become more committed to using the TWI methodology because they understand that it is based on solid theories, and they can now understand how and why it works.

Another approach to learning theory that is introduced is the learning styles model and new trainers are always eager to explore this part. Learning style is the way in which each learner begins to concentrate on, process, absorb, and retain new and difficult information.* The interaction of these elements occurs differently in everyone. Therefore, it is necessary to determine what is most likely to trigger each learner's concentration, how to maintain it, and how to respond to his or her natural processing style to produce long-term memory and retention.

Everyone has a learning style, and everyone has learning style strengths. It is easier to learn through strengths than it is to learn through weaknesses. Unfortunately, when instructors teach an entire group in exactly the same way, some learners are taught through their strengths; at the same time, others are taught through their weaknesses.

The model is based on the theory that

- Most individuals can learn.
- Different instructional environments, resources, and approaches respond to different learning style strengths.
- Everyone has strengths, but different people have different strengths.

* Dunn, R. and Dunn, K. 1978. *Teaching Students through Their Individual Learning Styles: A Practical Approach*. Reston, VA: Reston Publishing Company.

The Dunn and Dunn model* has four sensory modalities:

- *Auditory*: You learn by listening.
- *Visual*: You learn by seeing text (reading) or by seeing real or mental pictures.
- *Tactile*: You learn by using your hands and feet.
- *Kinesthetic*: You learn by using your body and senses.

People remember new and difficult information through different perceptual modalities. Although all instructors teach by talking, learning by listening is the most difficult way for most people to remember new and challenging information, particularly facts. On the other hand, in working environments like factories, it is common to "teach" people their jobs by having them watch an experienced person for as long as it takes until they "get it." But, they cannot capture the details or tricky parts of the tasks simply by watching and eventually wind up struggling and learning from their mistakes. In this part of the trainer development process, the trainers work with an exercise in which they discuss how they apply *all four* learning styles in the JI four-step instruction method by having the trainer "tell, show, and illustrate" the details of the job in step 2 and then having the learner actually perform the job in step 3. Finally, they make their own learning style profile to be more aware of their own style and to make sure they are able to adjust to individual learners.

Module 2: DiSC Introduction

We have chosen to work with the DiSC profile (see Chapter 5) as a psychometric tool for the job trainers. DiSC is a tool that lets the trainers improve on their self-knowledge and understanding of different behaviors. The theoretical model was first introduced by William Marston in his seminal 1928 book, *Emotions of Normal People*.† The purpose of introducing the trainers to DiSC is so they

- Understand their own personal style of behavior and how their behavior affects others
- Understand other people's styles of behavior
- Have a positive attitude of themselves, which generates confidence in others
- Adapt one's trainer style to other people's needs and situations

---

* Ibid.

† Marston, W. 1928. *Emotions of Normal People*. New York: Harcourt Brace and Company, London: Kegan Paul, Trench, Trubner & Co. Ltd.

Here, the trainers start to become conscious of their own preferred natural behavior style and learn how to be aware of other different styles when they are training. It is quite impressive how, after this module, the trainers can give examples of the behavior styles and patterns of their colleagues and the people they are going to train. DiSC is a user-friendly tool, and it is easy to work with.

This is followed by an individual feedback session during which each trainer receives thorough feedback and an overview of their own behavior style and pattern. We have a dialogue with them on what to focus on and develop as a trainer.

Module 3: Communication and Feedback (4 Hours)

The third module is communication and feedback; the trainers are introduced to the topic of communication, first more generally and then later in more detail as we relate it to the DiSC model to, again, make them more aware of their own preferences and how to adjust to the behavioral styles of others. They learn to read the DiSC styles of others by seeing videos of people acting within different verbal and nonverbal communication patterns. We also focus on how to give and receive feedback, which is an important discipline for trainers to master by practicing giving and receiving feedback in an appreciative way. Within the manufacturing areas, the appreciative approach to dealing with people is often overlooked. In several trainer teams, we have seen trainers become emotional when they try it out on their own "body and soul" to see the effect and the impact of this positive way of communicating with people.

Module 4: Training Challenges

The last module is on the training challenges they will encounter when working with difficult participants and how to handle these. Here, we also consider what training difficulties they fear the most. Reflecting on these, they try out those situations using role plays in small groups with opposite behavior styles. For example, a trainer who does not have a dominant behavior style will be paired in a role play with a dominant personality challenging the new method and process that has to be trained. Trainers are most nervous about training veteran employees who question why they must change the accustomed methods; here, we give the trainers a chance to try to overcome their fears.

Building Cultural Competencies

To build cultural competencies for the global roles in the training organization, we also provide cultural training and development activities. The mission is to be able to work effectively across cultures to create sustainable global collaboration and communication networks. This is one of the few courses that are conducted by external consultants. When we have a new global team start, we make sure the team is introduced to this global culture in which team members come from the different countries and cultures making up the LEGO team of companies. As an e-learning activity, they make cultural profiles of themselves to provide insights into knowing themselves better and then make these available to others in the group to leverage similarities and bridge gaps.

As we continue working in this cross-cultural environment, we are constantly building trainers with cultural competencies. Based on sessions at which we first introduced them to what culture is all about along with exercises to give them insight into how different we behave in different cultures, we are working on building consciousness of our personal reactions when we work with people from other cultures. This work is situational. Many of the situations during the workshops discussing misunderstandings based on cultural differences are based on actual situations encountered in the LEGO Group. Being open minded about dealing with these situations is one of the most difficult but important things for them to learn to do.

English Skills

For the people working globally in the training organization, the importance of language is essential to every aspect of their role and every interaction between the members of the global team. This means global job trainers, FAMTs, Learning Center consultants, and technical writers. This has been and continues to be a big challenge. Language is the foundation for the ability to communicate, form bonds, and create teamwork, and it is a precondition for good progress in the global process. Especially, the global job trainers are challenged to be able to function with sufficient English skills because they are coming directly from the production lines where English skill is not required or even desired, different from the other roles with candidates coming to these jobs with a more diverse educational background. We have had many discussions on whether to hire the global trainers with a higher educational background to ensure proficiency in English. But, we have realized that, even more than just English fluency, it is important for them to have knowledge of LEGO processes and to develop global trainers from our internal specialists on the floor who have a minimum 3 years of LEGO experience.

The way we develop their English skills is by intensifying our focus on language training; our Mexican site has also complemented the standard English language courses with additional conversational practice sessions. These sessions are driven by the technical writer or English teachers, who already, based on the skills profile for the role, have a solid educational background and experience with translation, interpretation, and communication in general. Here, they practice their English conversational skills on relevant issues while also preparing global workshop presentations to use as discussion tools. These English language classes have turned out to be a beneficial supplement to the overall trainer development program, and they are accelerating the language acquisition process, which is a long-term endeavor.

Evaluation Process

The final part of the trainer development process is evaluation (see again Figure 6.2). Before the trainers are ready to train the workforce, we evaluate whether they have mastered the JI method and the topics we have been working with in the Training and Learning Skills Course. The evaluation is based on an observation of a JI training session and the individual feedback that follows. There is also a current evaluation of the trainers through observation and feedback.

The entire process is supported by the HR Learning Center and by the managers and FAMTs. Figure 6.12 shows an overview of the evaluation process during which trainers are evaluated more intensively in the beginning and gradually less over time.

Learning Points: Building the Training Organization

As you have learned in this chapter, we are building the training organization by selecting the right trainers, developing them mainly through our own programs that we have created ourselves, and evaluating them continuously to ensure they deliver the training at the necessary quality level. But, we did not simply pull this process out of a hat. It took a lot of effort during the preparation stages, and we learned a lot during the pilot project, in which

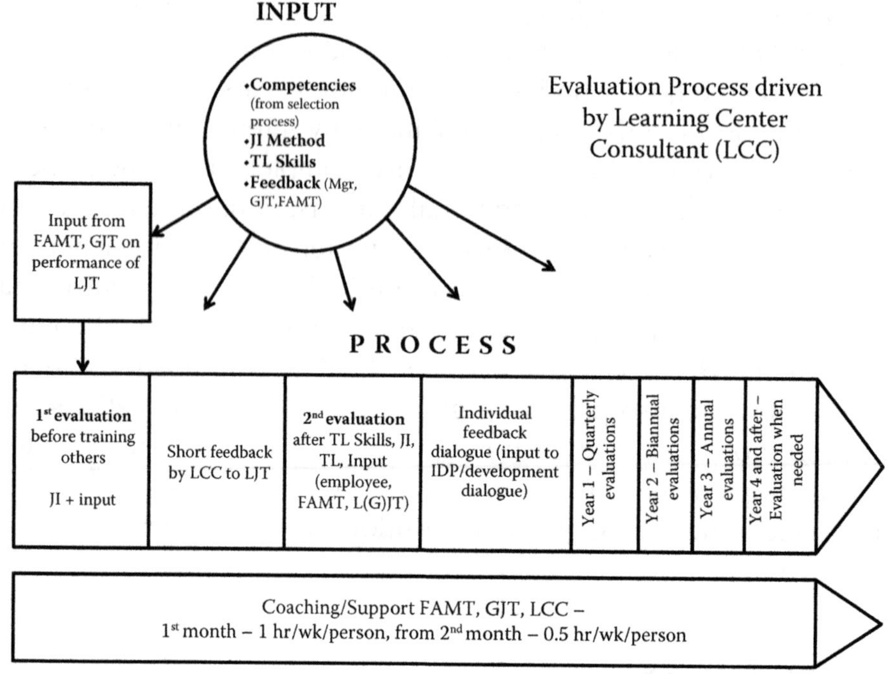

Figure 6.12 Evaluation and follow-up process.

we were able to try it out and make adjustments as needed. Here is a list of the key learning points we obtained from the whole endeavor:

- What works is developing concepts globally and then executing them locally. The way we have developed the concepts globally has created strong ownership among all participants, and all tools and processes are now anchored on all sites in a standardized way.
- Good investment in Job Instruction Trainer Development certifications and initial coaching from the TWI Institute were critical for ensuring quality of the instruction process and the Job Breakdown content and skills. It is important to learn to do it right from the beginning so motivation and trust in the method are not lost.
- Trainers are eager to learn and to be reflective in their roles and activities in the JI skill, feedback skill, leadership, and so on.
- The trainers are interested in obtaining insights into their own learning styles. Understanding behavioral types allows trainers to be able to recognize their own trainer style and adjust accordingly, thereby improving the quality of the training. One warehouse operator who has been working for the LEGO Group for more than 30 years said that she has

never learned so much about herself in her time with the company as she did in the 1-week Training and Learning Skills Course.

- We had invested a lot of time in stakeholder management at the general manager level but were lacking in activities and training for first-line leaders and management. This led to a lack of support in development activities and standardized work. It is difficult to understand and support a project if you are only introduced to it at a high level. It is important, then, to have these groups participate, for example, in the selection sessions and the 10-hour JI class.
- More training and coaching in leadership skills is necessary for the FAMTs to have them drive the local training organizations and support the global alignment process. It requires good facilitation and leadership skills to have trainers follow this vital direction, and our FAMTs struggled greatly in the startup.

Finally, to sum up what this work meant to all of us in the LEGO Group, here is a sample comment from one of our trainers that captures the feeling of those who went through this training:

> I have learned a lot. Now the jobs are taught step by step, where before it was the whole burger in one bite. I develop my ability to become a better trainer. It gives me more patience, and makes me grow as a person.

Enrique Quirino, Mexico, 2011

Chapter 7

Learning Tools and Methodologies

Introduction

In the previous chapter, we explained the development and content of the Global LEGO Training Organization that was created to address the needs of an expanding global workforce and tested and approved through the running of a pilot project in molding. Now, the pattern was set, and the final task would be to thoroughly embed this new system in the organization for use on a regular basis and become part of the culture of the company and how it went about its business. To do this, we would have to have tools and methodologies that the organization could use to leverage the new training organization we put into place. Moreover, we would need a system to drive the whole process.

This chapter explains the process we set up to execute effective training that enhances global learning and effective standard work. Furthermore, we elaborate on how we work with different learning methodologies depending on whether the training required is based on skills proficiency or knowledge acquisition. The process is started in collaboration with area leaders in terms of setting the scope of training and development activities by preparing a competency overview and a skills matrix so we know which processes and skills to use. These tools provide a common picture of what training is needed in the area and, in addition, are also used for evaluation of the workforce after the training and practice periods are completed.

The full process of systematic capability building is supported by the Learning Center consultants, who coach and guide the trainers.

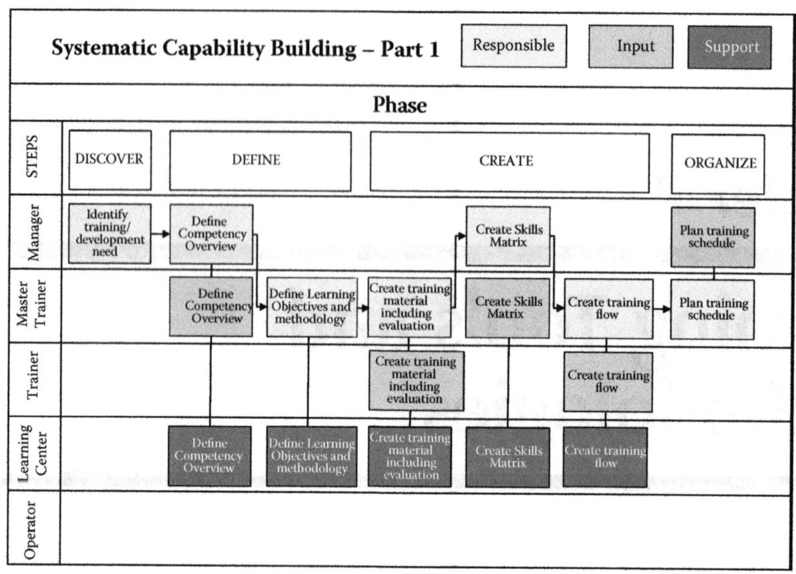

Figure 7.1 Systematic capability building process: Part 1.

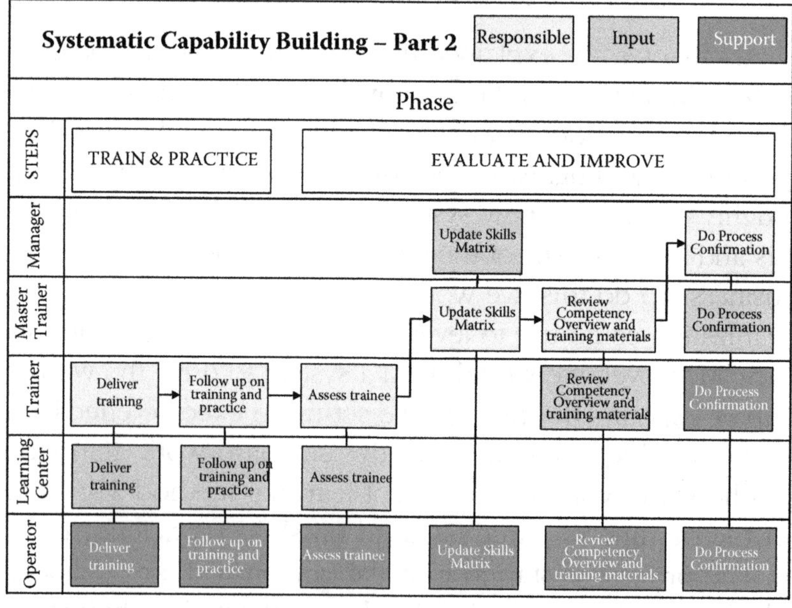

Figure 7.2 Systematic capability building process: Part 2.

The process model shown in Figures 7.1 and 7.2 provides an overview of the six different phases of the process: discover, define, create, organize, train and practice, and evaluate and improve. The rest of the chapter follows this process, and the tools and methodologies we use, from start to end.

Competency Overview and Learning Objectives

When starting in an area where there is a need for knowledge transfer or a new process is to be developed, the first step is to clarify the scope of work for that specific process or area. Our primary tool in defining the scope of competencies is the competency overview (see Figure 7.3).

The function of this tool is to create a clear picture of just what competencies will be tied to the training material to ensure that we, in answering the training need, will meet the scope required. Based on the way we work with competencies in that area, we will consider the *knowledge*, *skills*, and *behavior* used in the work to create an overview of the competencies required within that specific area or function. The overview is used as an initial tool to support a leader in prioritizing training and to give a full picture of how much training material is needed for each competency. By having a clear overview of the competencies, the leader is also able to prioritize and focus on the core and critical competencies first.

The competency overview structures how to set up the training material and which training method to apply for each competency. Each competency is defined by breaking it down into overall purpose and learning objectives as well as time required for training. All this is done with support from the Learning Center consultant and will give the leader and the trainer a clear direction for the structure and content of each training activity. We should emphasize that this tool must be made together with the leader of the area because we are defining the scope of the process for which the leader is in charge. In only a few hours, they can create this clear picture of what the area is capable of, and this gives them a good overview of what to do in terms of defining priorities, where to start, and so on.

The first task is to define the *heading* of the competency overview. The heading explains the full scope of the competence breakdown. It can be a whole area or department, a specific role in a department or area, or a specific type of competency (knowledge, skill, or behavior) you want to map across departments or areas.

The next step is to add *topic titles*. There are several ways of doing this. You can start by asking one employee to specify his or her main tasks and then ask what competencies the employee needs to perform these tasks. These competencies can be inserted as topic titles or headlines. As you ask more employees doing the same job, a common picture will begin to emerge. If the area you are mapping is characterized by numerous specialists with many different tasks, you can let the topic titles be

Competence Breakdown Molding Setter Basic Knowlegde

General Introduction Day	Training Method	IN/EX	Local Area Introduction	Training Method	IN/EX	Machine Knowledge	Training Method	IN/EX	Basic Material Knowledge	Training Method	IN/EX
General Safety Rules	C		Welcome	C		Machine Safety	C/S		Material Safety	C	
Quality Intro	C		Materials at LEGO	C		Ejector Types	C		Material Chemistry	C	
Common Words Glossary	C		Basic Molding Training	C		8 Machine Main Parts	C		Plastic Material Handling	C	
Lean Introduction	C		Movie/Tour	C		Direct and Indirect Setup	C		Test	C	
[Topic]			[Topic]			How to Tight Bolts	S		[Topic]		
[Topic]			[Topic]			Nozzle Types	C		[Topic]		
[Topic]			[Topic]			Handling of Nozzles	S		[Topic]		
[Topic]			[Topic]			B Button	S		[Topic]		
[Topic]			[Topic]			Alarm Messages	S		[Topic]		
[Topic]			[Topic]			Alarm Lights	S		[Topic]		
[Topic]			[Topic]			Adjustment of Flip flop	S		[Topic]		
[Topic]			[Topic]			Emptying Barrel	S		[Topic]		
[Topic]			[Topic]			Machine Modes	S		[Topic]		
[Topic]			[Topic]			Control Panel	S/C		[Topic]		
[Topic]			[Topic]			Test	C		[Topic]		
Sequence	?		Sequence	?		Sequence	?		Sequence	?	

Injection Molding Process Knowledge	Training Method	IN/EX	Type of Molds	Training Method	IN/EX	Equipment	Training Method	IN/EX	Maintenance	Training Method	IN/EX
Pressure Methods	C		Safe Mold Handling	C/S		Safe Equipment Handling	C/S		Safe Maintenance	C/S	
Decompression	C		LEGO HK ver. 1	C		Robots	S		Soft Parts	S	
Machine Cycle	C		LEGO HK ver. 2	C		Grinders	S		Ejector Protector	S	
Parameters (speed, pressure, parameter)	C		LEGO HK ver. 3	C		Mold Temperature Controller	S		Clean Mold	S	
Back Pressure	C		LEGO HKN	C		Conveyors	S		Mold Lubrication	S	
Mold Protection	C		2 Plates	C		Part Controller	S		Clean Machine	S	
Injection Time Monitoring	C		3 Plates	C		Safety Injector Cable	S		Machine Lubrication	S	
Material Cushion Monitoring	C		HD Molds	C		Material Blenders	S		Change Wear Parts	S	
Flow Rate Regulation	C		Core Pull Molds	C		Cooling Pipe	S		Test	C	
Test	C		2 & 3 Components mold	C		Cooling Drum	S		[Topic]		
[Topic]			Co Injection Mold	C		Preheater	S		[Topic]		
[Topic]			Test	C		Crane	S		[Topic]		
[Topic]			[Topic]			Hand Tools	S		[Topic]		
[Topic]			[Topic]			Test	C		[Topic]		
[Topic]			[Topic]			[Topic]			[Topic]		
Sequence	?		Sequence	?		Sequence	?		Sequence	?	

Legend			
Structured shopfloor training	S	IT training / eLearning	I
Classroom training	C	Mentor-based training	M
Internal training	IN	External training	EX

Figure 7.3 Competency overview example.

more task or process oriented. It is a good idea to apply colors to classify similar competencies or to separate them. Examples of groups of similarities might be things like basic material knowledge, type of molds, machine knowledge, information technology (IT) systems, documentation, equipment, maintenance, manual measuring, calibration, software, and so on.

After adding in the topics, next you define the type of *training method*. This part of filling out the competency overview sometimes has to be revised several times before the right learning objectives or training methods are found. It depends on the content of the competency and the learning process. For example, if it has to do with understanding theory or knowledge acquisition, classroom training might be the right training method. Different abbreviations for the differing training methods and classifications are used: C = classroom training (knowledge), S = shop floor training (skills), E = e-learning (skills and knowledge). Sometimes, a blended learning approach is needed. These three types of instruction are explained at the end of this chapter.

There is always great disparity in how much training material is currently available in any new area we go into to help with training. Sometimes, there are many good guidelines or standard operating procedures (SOPs) already in existence; at other times, we start from zero with no documentation available. We should use the material already existing as a starting point because much knowledge is provided here that we will need to access. Furthermore, it is important to value the efforts of the current trainers or the organization overall because they created these resources; with value, they will not become defensive and antagonistic to our efforts. We will need their help and participation as we move forward.

The functional area master trainer (FAMT) works together with the trainers when starting the process of training material setup; they will use the competency overview to review the overall purpose, learning objectives, and time required for training. While setting up the training material, the competency overview is often adjusted and learning objectives added.

Learning Objectives

For each topic, the *overall purpose* and *learning objectives* need to be defined. The overall purpose is a short description of what exactly we are trying to accomplish. The learning objectives are expected outcomes of the training and are more concrete and clearly stated. What should we expect from the employee after successful training? The purpose of defining learning objectives before creating a training activity is to set a direction and make clear the goal that is to be reached. Once set, all decisions concerning constructing and carrying out the training activity must be measured in relation to the learning objectives. When setting learning objectives, we use the following questions to define the objectives:

- What will a learner be able to do as a result of the training?
- What is the focus? What do we want to enable the target group to achieve with the new knowledge?

When we evaluate the training, we always do it based on the defined learning objectives. During the process of filling out the competency overview, the trainers go back and forth to align the topics and topic titles in an iterative process as more details emerge. An example of the overall purpose and learning objectives for basic material knowledge training is shown in Table 7.1.

Table 7.1 Example of Learning Objectives

Basic Material Knowledge				
Topics	*Training Method*	*Overall Purpose*	*Learning Objectives*	*Time*
Material safety	C	Understanding of the basic knowledge for plastic material and the process from resin to LEGO elements	1. From raw material (oil/gas) to plastic - molecular chains 2. Difference between thermosets and thermoplastics 3. Difference between semicrystalline and amorphous	
Material chemistry	C	Rules for handling material and understanding the color code system	1. What to do with material without identification 2. What to do with contaminated material 3. What to do with regrind material 4. What to do with material after changeover	
Plastic material handling	C	Ability to handle plastic material safely	1. Why we use safety gloves 2. How to ensure deporting of melted plastic in a safe place 3. What to do if you have an accident with hot plastic	

When a trainer starts working with the training topics defined in the competency overview, it could be the first time he or she ever needed to transfer practical knowledge into documentation. The ability to do the job is different from understanding that job and writing down what you should be able to do to perform it. Here, support from the Learning Center consultant is important; trainers are given specific examples of other learning objectives from other areas. The Learning Center consultant guides and coaches the trainers through the process, especially in the beginning, as the trainers are usually unfamiliar with how to think theoretically about the job. Many activities are tacit knowledge for them, and they do them automatically without being aware of what they are doing.

After having defined the learning objectives, the *learning method* should then be defined. Again, the Learning Center consultant is guiding them, first by helping them to define whether the task is knowledge, skill, or behavior. Based on this, they decide if it makes sense to set up a training program for classroom training, which communicates knowledge to a group of people; one-on-one training, for which skills are trained using the Training Within Industry (TWI) Job Instruction (JI) methodology; e-learning, by which they learn on their own with the use of a computer; or a blend of these. Figure 7.4 shows a depiction of the different learning methodologies.

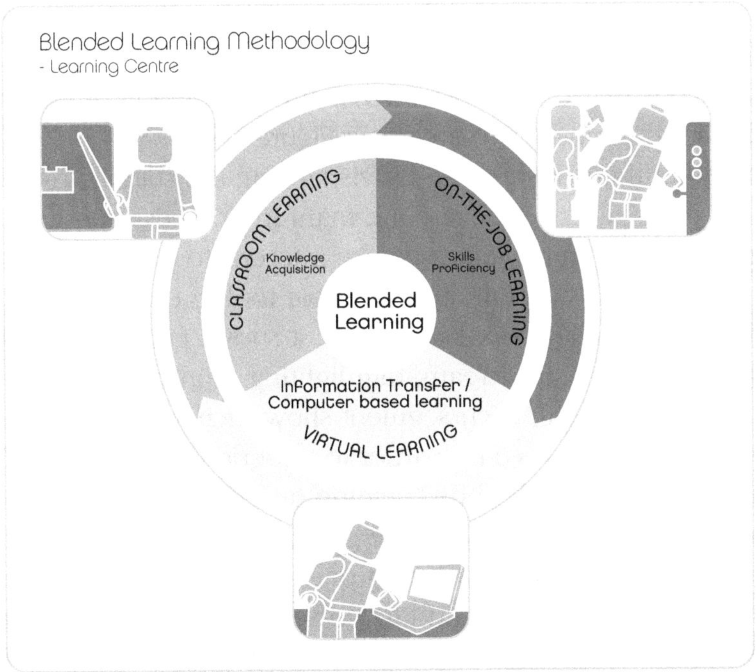

Figure 7.4 Learning methodologies.

The advantages of setting up the competency overview and the learning objectives are that they give a clear picture of what training material should be set up; furthermore, this provides the leader and the team with a common picture of how to start up and proceed with defining the training material. The challenge of this process is that it requires much guidance and resources from the Learning Center, as it is generally difficult for new trainers to define learning objectives and learning methodologies. But, with consistent guidance, these same trainers gain experience and become better and better at handling new training demands in the future.

Skills Matrix

Another tool being used for mapping of skills and evaluation of the workforce is the *skills matrix*, which is commonly used in Lean activities for shop floor management and for training and development purposes. The overall objective of this tool is to avoid giving employees jobs they do not have the skills to perform. The skills matrix provides a visual and transparent way of matching needs with available skills and indicates training requirements for filling in the gaps. Furthermore, it gives people an understanding of what requirements are needed for a new task or role and then lays out the individual names of the people in a work group and the main skills required to perform the jobs in that area. The skills matrix is displayed visually and openly on the local information boards in the production area (see Figure 7.5).

The purpose of the skills matrix, by visualizing the status of skills by employee and by area, is to facilitate the manning process and to create flexibility in each area. In addition, we use it to create training schedules identically in line with the TWI Training Timetable tool to ensure that we have the needed skills in each group. In addition, it shows group targets for each individual and for each work stream, highlighting training requirements, and the quadrants (of the pie) in each segment show increasing skill levels as the learner progresses from novice to learner to experienced operator. Each level of skill can be further described by "helping questions" (see next section).

It can be a challenge to handle large groups in one skills matrix and to show if people have skills in more than one area, so it is important to keep the matrix active and up to date and to do so when processes or machines are changed or changes in the workforce take place. When using the skills matrix, you let people know who they can turn to for support because it shows who else is proficient in any one task or area. Moreover, it helps

LEGO **Skills Matrix**

Last Update: 30.05.13 | Area/Role: Spareparts and moulds | Responsible Person: Jim

Skill / NAME	ID	Basic knowledge/skill	CAD – tool maker	CAM – 3 axis	CNG Milling 3-akset	CAM 5-akset	CNG Milling 5 akset	CNG Turning	D Cam	Wirecutting Progress (P&T)	Wirecutting Cut20 (LOM/KOM)	Spark Erosion – Hyper Spark (P&T)	Spark Erosion – Form 20 (LOM/KOM/TO)	Spark Erosion – ZK (P&T)	Engraving	Assembling, testing and adjustment	Grinding	Polishing	Quality and measurement skills	Planning skills		Number of skills trained in															
	No.	1	2	3	4	5	6	7	8	9	10	11	12	13	14	15	16	17	18	19	20	1	2	3	4	5	6	7	8	9	10	11	12	13	14	15	16
John		◑	◑	◑	◑	○	○	○	○	○	○	○	○	○	○	○	◔	○	◑	◔	○																
Pete		◑	◑	◑	◑	○	○	○	○	○	○	○	○	○	○	○	◑	○	◑	◔	○																
Rich		◑	◑	◑	◔	○	○	○	○	○	○	○	○	○	○	○	◔	○	◔	◔	○																
Joe		◔	○	○	○	○	○	○	◑	○	◔	○	◑	○	○	○	○	○	◔	◔	○																
Vince		◔	○	○	○	○	○	○	◔	○	◔	○	○	○	○	○	○	○	◔	◔	○																
Ed		◕	○	○	○	○	○	○	◕	○	◕	○	◑	○	○	○	○	○	◕	◔	○																
		○	○	○	○	○	○	○	○	○	○	○	○	○	○	○	○	○	○	○	○																
		○	○	○	○	○	○	○	○	○	○	○	○	○	○	○	○	○	○	○	○																
		○	○	○	○	○	○	○	○	○	○	○	○	○	○	○	○	○	○	○	○																

Figure 7.5 Skills matrix.

people to see their own need for development. Management can obtain important information quickly to support the manning process. Finally, the skills matrix can point out and make clear what minimum skill requirements are needed for each role. This is a big advantage when coordinating between production sites as it creates a common picture of what is required, the target, for each level in the different roles.

Helping Questions

The skills matrix describes the skill level of each worker performing each task in the form of a circle (or pie) in each segment of the grid; the more it is filled in, the higher the level of skill of the operator will be. We have created a list of helping questions that define further what is needed to advance from one skill level to the next level as indicated by the filling in of the four quadrants of the circle (pie) in each segment of the matrix. This is extremely helpful when defining those skill levels to have a shared and coherent picture of what it takes to achieve that skill level, especially when working globally across production sites. This makes it possible to base development plans and, for instance, promotions on actual skill levels in an objective way and not on subjective dialogues and opinions.

This tool makes it much more transparent for employees to see what is required of them to advance. Another benefit is that we can standardize

Table 7.2 Example of Helping Questions

Helping questions					
P	Skill	Level of skill			
4	D-CAM	Completion of training task part I D-CAM basis Export Step 214 part from NX Import Step 214 to D-CAM Completion of IT structure training Completion of 10 practical operational tasks (2 months) Evaluation task 8	Completion of training task part II Sketch Completion of 15-20 practical operational tasks (during 3-6 months). Evaluation task 24	Completion of training task part III Programming "Solid" Completion of 15-20 practical operational tasks (during 5-7 months) Evaluation task 27 (collar ring)	Completion of training tasks: Completion of practical operational tasks in a period of 1 year (in total 2 years of practical experience) Evaluation task 30: Test brick creation and solids
5	Wirecutting Cut20	Machine introduction and safety (1 week) Completion of training tasks part I Completion of 10 practical operational tasks (2 months) and support on other operational tasks Evaluation task 8	Completion of training task part II Sketch Completion of 15-20 practical operational tasks (during 3-6 months). Evaluation task 24	Completion of training task part III Programming "Solid" Completion of 15-20 practical operational tasks (during 5-7 months) Evaluation task 27 (collar ring)	Completion of training task part IV Completion of practical operational tasks in a period of 1 year (in total 2 years of practical experience) Evaluation task 30: Test brick creation and solids

the skill levels of the employees, providing a higher level of flexibility across global production sites, because we know exactly what is required to develop the people and their skills. As you see from Table 7.2, each level of skill contains formal training, practice periods, and tasks, all supported by follow-up and coaching, so the development of skills is based on the 70-20-10 learning principle we talked about in Chapter 6. We cannot emphasize enough that you do not acquire a skill simply by attending a formal training session but through actual practice and experience.

Individual Training Plan

As we saw in the previous section, the skills matrix tracks closely to the TWI Training Timetable, which is a training plan to help us determine *who* we should train, in *which job*, and by *what date*. When we can answer these three questions, then we have a training plan; it is important to plan the training by applying several different approaches. For example, we can look for and plan to fill urgent training needs as they come up in the daily work of the department or, in another application, plan the development of a brand new person from teaching the simple tasks first building up to the more complex and difficult ones until the person has learned and can perform the entire scope of the position.

The training plan describes, in detail, which type of training will take place and when the training will occur. This is important to ensure that the trainer has the time to prepare for the training and that the trainee knows when the training will take place. Planning of the training must be aligned with the daily activities and operations of the department. The training plan is also used to create an overview of the lead time needed for completing the required training or the time needed to build up a new position. In the following example, shown in Figure 7.6, you see the timing of the training for a wire cutter technician. By setting up this visual plan, it becomes clear how much time it will take to train, allow practice, and evaluate the new person.

Evaluation

The evaluation process lets us ask the critical question: Have the goals and objectives of the training and the competencies been met? The purpose of the evaluation is to assess how well individual achievement has satisfied needed requirements; it provides information that can be used to improve the competency level of the worker and document accomplishments or failures. So, after the training is completed, the trainee is evaluated to see whether he or she has succeeded in obtaining the expected skills, knowledge, and behavior according to the training program and the defined learning objectives. The evaluation can provide feedback and motivation for continued improvement of the trainees, trainers, and leaders. This means that by evaluating the trainees, it also gives feedback to the trainers regarding whether the training they gave was successful or whether adjustments are needed.

The evaluation can be theoretical, practical, or a combination of both. The theoretical evaluation consists of questions, like a test, based on knowledge provided through classroom training. The evaluation is matched against the learning objectives defined for the training activity. The practical evaluation, on the other hand, is based on a specific task to be performed, so it assesses whether the needed skills were acquired from the training sessions, follow-up support, and the important *practice* that is required to build these skills. Every training period and all milestone activities should be evaluated; it is vital that the trainers have the time to perform follow-up on training periodically. Figure 7.7 shows the evaluation form used in this process.

By using the helping questions in the skills matrix, we define the development process for each level of skill and therefore what needs to

Name				Calendar week	1					2					3					4	5	6	7
Role: Wirecutter				Training day No.	1	2	3	4	5	1	2	3	4	5	1	2	3	4	5	All	All	All	All
	Provide	Required	Actual	Gap																			
Training Competencies																							
Repeat training, Support and daily tasks																							
Training activities Wirecutting - Intro to training		●	○	100	↔																		
Training activities Wirecutting - IT structure		●	○	100	↔																		
Training activities Wirecutting - Work planning		●	○	100	↔																		
Training activities Wirecutting - D-CAM Program knowledge (Ikon guide)		●	○	100		↔																	
D-CAM Theory - 30,1 D-CAM Two Point Line		●	○	100			↔																
D-CAM Theory - 30,2 D-CAM Line given Angle		●	○	100			↔																
D-CAM Theory - 30,3 D-CAM Parallel		●	○	100				↔															
D-CAM Theory - 30,4 D-CAM Circle Given Center		●	○	100				↔															
D-CAM Theory - 30,5 D-CAM Three Point Circle		●	○	100					↔														
D-CAM Theory - 30,6 D-CAM Circle Given Radius		●	○	100						↔													
D-CAM Theory - 30,7 D-CAM Square		●	○	100						↔													
D-CAM Theory - 4. Export stepfile 214 LOM Training		●	○	100							↔												
D-CAM Theory - 5,1 Import stepfile LOM træning		●	○	100							↔												
D-CAM Theory - 5,2 Generate program LOM training		●	○	100								↔											
D-CAM Theory - 6,1 Data transfer		●	○	100									↔										
Wirecut Theory - 7,1 Preparation of machine supply LOM Training		●	○	100									↔										
Wirecut Theory - 7,2 Preparation of machine alignment LOM training		●	○	100										↔									
Wirecut Theory - 8,1 Test of element		●	○	100											↔								
Wirecut Theory - 8,2 Fixture of plate		●	○	100											↔								
Wirecut Theory - 8,3 Fixture of cavity		●	○	100												↔							
Wirecut Theory - 9,1 Check - Dial gauge		●	○	100												↔							
Wirecut Theory - 9,2 Choose misalignment cycle LOM Training		●	○	100													↔						
Wirecut Theory - 9,3 Data transmission to program		●	○	100														↔					
Wirecut Theory - 10,1 Check program		●	○	100														↔					
Wirecut Theory - 10,2 Start program		●	○	100															↔				
Summary of first training period		●	○	100																			

Figure 7.6 Example of a training plan.

LEGO Evaluation

CAM

Created by: Christoffer Havsgaard, Hans-Jørn Kristensen	**Authorized by:** Claus Morten Larsen	**Issue date:** 24-02-2012

Employee name: Joe Smith

CAM basic 1 Electrode Manufacturing

Task 1

Preparation of program
All OK

Cavity Mill
Used correct feed and speed and cut levels; Corner smooth not used.

Z Level
Used correct feed and speed and cut levels; Corner smooth not used (better for finish)

Face Mill
Didn't use it

Contour Area
Used well but have some recommendations

Planer
All OK

Planer Text
Didn't use it

Milling Strategy
Followed correct sequence stock, could be better if you use the correct path for rough semi finish

Conclusion: Retraining in Z level, contour areas, planer text and milling strategy.

Date of Evaluation:

Trainer

Figure 7.7 Example of a training evaluation form.

be learned at each stage. This means that it defines which training tasks should be completed, by when, the content of the practice period, and finally, which evaluation task should be performed before the learner can proceed to the next skill level. The performance of the task is thoroughly evaluated, and an important part of this evaluation is the individual feedback to the trainee. It is finally decided by the trainer and the leader of the area whether the trainee is ready to proceed to the next skill level. The biggest challenge here is that it is time consuming to generate the documentation and the training material, but with the strong support of the Learning Center consultant we can ensure an appreciative and positive experience for the trainee.

Learning Methodologies

As pointed out previously in this chapter, we use a blended learning approach of various types of training and learning techniques to achieve the most effective performance gains in the most efficient manner. The components include a combination of classroom training, on-the-job (shop floor) training, and virtual training and learning activities (see Figure 7.4).

On-the-Job Training: Job Instruction

We have described throughout this book the implementation and use of the TWI JI method and why we chose this method to embody the core skill of our global learning organization. It provides us with a standardized method of on-the-job training and a way to break down jobs for training that can be easily translated and transferred to other production sites in countries around the world. This tool provides us with a common language and methodology for training jobs so that any LEGO employee, no matter where in the world the employee works for the company, will receive the same instruction and be able to perform the job in the same way as any counterpart in LEGO production sites around the globe. Moreover, the program is a powerful instruction method that motivates employees to follow standard methods so that the overall goals of continuous improvement and Lean systems can be achieved.

We could not describe the many details and nuances of the TWI JI method in this book, but they can be found in *The TWI Workbook, Essential Skills for Supervisors* (P. Graupp and R.J. Wrona, Productivity Press, New York, 2006). It is sufficient to say that this method is a one-on-one, hands-on approach to job training focused on skills instruction. In other words, this is on-the-job training (OJT); learners actually witness and perform the jobs they are learning during an intensive session with trainers. In Step 2 of the four-step method, present the operation, the trainer actually performs the job several times for the learner to see while giving additional information in each round of the demonstration: first the Important Steps, then the Key Points for each of these steps, then the Reasons for each of the Key Points. After that, in Step 3, the try-out performance, the learner then performs the job repeatedly, each time repeating those same Important Steps, Key Points, and Reasons for Key Points as the learner performs the job. The learning occurs through this repetition and practice, which is done in the production area, right on the job when possible, but always near the job and in realistic working conditions when that is not possible.

When it comes to learning to perform jobs, nothing can substitute for actually getting your hands on the real jobs and practicing them as you do the work. This is essential when learning a skill because we learn skills through practice and repetition. The advantage of using the JI method was it enabled all LEGO employees to be taught in the same disciplined way and to hear and understand the same procedure on how to do the job. In this way, everyone practiced it in the identical manner, and we could obtain and enforce standard work.

However, there are many aspects to doing any job that entail knowledge in addition to the skills needed to perform the work. Here, it would not be efficient to teach this knowledge in a one-on-one, on-the-job setting. So, as we will explain, we also had to give classroom training to teach those aspects of the work so that our OJT would not become bogged down in details that would ultimately drive the training out of the working day. It did take time and was labor intensive to perform proper JI, so we needed to make sure the training was as efficient and effective as possible.

Challenges to Shop Floor Training

Preparing the Job Breakdowns and other training materials can be a long and tedious process, so it is important to support the trainer teams in carrying out this preparation work as well as in the teaching of those very jobs. When confirming whether the Job Breakdown will work or not, it is important to go out on the shop floor and try it out by teaching the job using the breakdown you have made and making adjustments as needed. So the work of training, and the total time it takes to do it, is not just the time spent with the learner teaching the job. It must include preparation time because that is critical to the success of the training.

LEGO	Disconnect Mold Temperature Controller Job Breakdown Sheet		
Created by: István N., László Zs., Bent J., Tommy H., Gustavo O., Alberto R.		**Document Owner:** Klaus A. N., Marien P. V., Flemming L. K.	**Date of Implementation:** 06.04.2012. **Date of Change:** **Estimated Training Time:**
Parts:			
Tools & Materials:			
Nr.	**Important Steps** A logical segment of the operation when something happens to advance the work.	**Key Points** Anything in a step that might — 1. Make or break the job 2. Injure the worker 3. Make the work easier to do; i.e., "knack," "trick," special timing, bit of special information	**Reasons for Key Points** Reasons for the key points
0	Common key point	1. Follow general safety rules	
1	Close flow meter no. 4	1. Closing water battery 2. Closing both valves 3. Closing them finger-tight	1. Close down flow meter 2. Releasing the water 3. Not to damage the threads
2	Turn off and release pressure (LOM, NYI)	1. Changing to vacuum mode	1. Risk of water on the floor
3	Turn off the power (BLL)		
4	Disconnect all hoses	1. Pulling the security cylinder 2. Placing the hoses up off the floor	1. It won't come off 2. So nobody falls over the hoses
5	Disconnect the Interface-connector and connect th blind-connector	1. Loosening the safety-lock 2. Grasping the connector and pulling 3. Locking the safety-lock	1. Can´t be disconnected 2. You will pull the cable out of the connector 3. Can fall out and the machine can´t run
6	Disconnect the Electric-connector	1. Lifting safety-lock 2. Grasping the connector and pulling	1. Can´t be disconnected 2. You will pull the cable out of the connector
7	Remove the Mold temperature controller	1. Wrapping the cables around the holders	1. So nobody falls over the cables

Figure 7.8 Job Breakdown showing differences in procedure by location. LOM, Mexico plant; NYI, Hungary plant; BLL, Denmark plant.

Although we have stated repeatedly that the purpose of implementing the JI training was to create standard work processes that would be performed identically in the different sites around the world, there are still situations for which there will be differences in the procedures because of environmental conditions, differences in machinery, differences in working rules and behaviors, and so on. Figure 7.8 shows a partial example of a Job Breakdown developed from a global perspective; you can see some differences for the different sites (indicated by parentheses with the site abbreviations, e.g., NYI for Hungary and LOM for Mexico). This means that we are not always working 100% according to the same standard; however, we will still be following the same principles for doing the job. In other words, there may be differences in procedure due to differences in machinery, conditions, and the like, but all sites will be following the same fundamental principles of how we want the job to be done.

Classroom Training

Classroom training is defined as standard instructor-led, face-to-face training in a classroom. The training is conducted by the job trainers. The purpose of classroom training is to capture and leverage the knowledge of the workforce within specific topics where it is possible to train a group of people to obtain that knowledge (i.e., material knowledge). We use different techniques, methods, activities, and training aids available to create and present memorable, meaningful, and successful classroom training sessions. We are using a variety of techniques to reach different kinds of learners and to achieve the learning objectives for specific target groups in the workforce.

Classroom Training in Material Knowledge

To give an example of one of our classroom training sessions on the topic of material knowledge for molding employees, our functional area master trainer Flemming Tiro Lund has designed a course based on his many years of experience with materials in the plastics industry. The objective of the training is that employees understand the properties of plastics and how to handle the materials to avoid stressing the material, which could result in defects, errors, and rejects. According to Flemming: "The employees do not need to become chemical experts with extensive knowledge in molecular structures and formulas, but the knowledge should be delivered in a way that everyone understands and with a high degree of practical exercises and relevant examples."

The training session starts with a short, basic theoretical introduction to the different kinds of plastics that we work with in production. As most of the problems in production occur from lack of knowledge in the handling of plastics, it is essential that workers understand the effect of exposing the plastics to certain temperatures, heating speeds, and so on. To develop this understanding, the classroom activities include some relevant "hands-on" experiences and exercises. So, after learning the basic theory, they make a visit to the shop floor and to a molding machine, where Flemming demonstrates how the material runs through the machine and what happens when the material remains too long in the machine. He also demonstrates for them what is happening when plastic material is stretched. This is done to show the process of transforming from a granulate to the ideal condition for the material to be injected into the mold.

Next, he takes them back into the classroom for a practical exercise. Flemming explains one of his exercises as follows:

> We always try out a practical exercise where I explain that the material can move and be influenced by many factors. I hand out a piece of chewing gum to everyone in the group. I explain that they should imagine that it is a piece of plastic material, and notice that, at the start, it is hard and stiff until the surface has been broken and it becomes easier to chew. They end up finding the good condition of the material by chewing. Then we put in an additive (wax) and this is done by giving everyone a little piece of chocolate to chew on just as additives, like wax, are added to the plastics. They feel suddenly the condition of the chewing gum changing into a more fluid condition and they can feel how the material separates and falls apart. We relate this experience to what they saw at the molding machine and what happens when you change the temperature, the time heating the material, etc. which make it possible to work even faster with the plastics without destroying its molecular structure. This is one way of trying to give the participants insight into the fact that the molecular chains in the material are alive.

Another exercise is with a plastic cup, in which the molecular chains are stretched in the length of the cup. The cup is heated to a specific temperature, and the cup slowly collapses and melts into a flat brick as the molecular chains are allowed to relax. This process is demonstrated in a minioven with a glass door that allows the participants to follow the process. This again provides insight into the dynamic properties of the plastic material, and that the way we handle it is important for the process and the output.

In the classroom training, the combination of theory and practice ensures a dynamic learning process for the people in production. At the same time, it creates an essential foundation of understanding that enables people to perform problem solving in production. It also provides production people with a good personal experience and engages them in a direct "hands-on" experience and focused reflection to increase their knowledge and develop skills. The advantages of classroom training are that it allows us to teach employees in a safe, quiet environment, away from the noise and pressures of the work area, which gives us an opportunity to have good group dialogue and interaction. In addition, we can train more people at the same time, which requires fewer resources. As group interaction enhances learning, the employees learn from one another as well as from the trainer.

One of the challenges to classroom training is that it can be unfamiliar and uncomfortable for new job trainers, who do not have experience leading large training groups, to stand in front of a group and facilitate sessions like this. When we started the pilot project, only the Denmark facility was conducting this kind of training on material knowledge. Now, all the sites have been trained in train-the-trainers sessions, and the trainers are conducting this class successfully.

e-Learning

In the LEGO Group, e-learning is an approach to learning by which the acquisition of new knowledge and skills is facilitated via ICT (information and communication technologies). We want to believe that e-learning is "something that happens and not something you take." For example, e-learning could happen when an employee uses a laptop to access an online interactive quick guide to solve a given problem on the job. There are many ways to use ICT to enable learning, and we call these e-learning methods.

An essential training activity is our IT training in programs like CAD/CAM (computer-aided design/computer-aided manufacturing), which is used for spare part and mold design and manufacturing. Here, we combine software recordings with face-to-face training as a way of demonstrating how to perform a given task with the program. After viewing the demonstration, which includes the Important Steps, Key Points, and the Reasons for Key Points (JI), the trainee does the job with the software recording as the "helping document." By using this method, the trainee is working more independently and is always able to use the software recordings at a later time for review. We have combined e-learning with the possibility of working more independently after the training by being able to return to the training program whenever needed.

So far, we have only used e-learning methods on a small scale, but there is no doubt that web technologies and various media platforms will play a greater role in the way we will support and enhance learning in the future—both in our formal training setup and in the workplace with our employees.

The purpose of e-learning is to provide a convenient, time-efficient, and flexible way of learning. What would create a real learning advantage would be a convenient e-learning portal that provides cost-effective and time-efficient methods for users to gain skills and knowledge whenever and wherever needed. We see this not only as an online self-paced training methodology but also as an enabler for people to learn anytime and anywhere. In the future, we see great potential in developing asynchronous learning activities for, say, molding operators to improve their skills in trouble shooting, which is a skill that is difficult to acquire and to gain experience in. We have many skillful molding specialists with much tacit knowledge in troubleshooting, and by applying their knowledge and experience in a virtual troubleshooting platform by means of text and images and communicating all this through a tablet computer platform, we could make the learning process more efficient. In the future, the intention is to combine the tablet computer platform with the TWI training. When a problem is identified, several solutions are possible to solve the problem. By having been trained in all possible solutions of known problems through the JI method, we can create a strong platform for solving problems in production.

Another area for intensified focus on e-learning is in setting up more interactive learning platforms for our toolmaker and plastic maker apprentices. From a global perspective, the main deliverable is a virtual learning portal that is both personal for the individual apprentices and social for the group of new and existing toolmaker apprentices. Thus, the portal will provide a personal overview of the education (content, structure, and progress) and easy access to best-practice material (PowerPoint, video, e-learning modules, etc.) in topics like drawing fundamentals or work planning. The topics will be supported and guided by virtual mentors in a professional online community (trainers being specialists in these topics) through both synchronous and asynchronous learning activities. In addition, a global apprentice community will be created with networking activities, and preparation is being made for a global exchange of apprentices.

We see the e-learning approach as beneficial to our future when combined with the existing high degree of on-site and face-to-face training in practical shop floor skills. By implementing the e-learning approach to learning and a truly blended learning perspective, we can, to a higher degree, ensure that e-learning is "not something you take, but something that happens."

ROLLOUT

Chapter 8

Sustaining the Effort and Growing the Future

> The benefits I see from the training organization/concept are that we have dedicated trainers (global and local) on each site for each area meaning they have it as a part of their job description and the time is allocated for training. They are known to the rest of the organization as the trainers and they are coordinating changes to the training material globally, so we have one method and one training material for each job/function.
>
> Besides this, I think that this has made our reaction time on getting new employees in place for new jobs shorter and aligned and we have the possibility to get rid of old habits in doing the different jobs. In addition, it also makes it possible to make real assessments of the competency level of each employee as we have a breakdown to compare with.
>
> **Senior Director, Molding Design and Implementation**

From Pilot Project to Program Management

After we closed the pilot project, we went, in just a few months, from a narrowly defined project to 22 diverse projects/initiatives spread throughout the international plants. Even while the pilot project was running, we had received several requests from other departments in LEGO Operations about obtaining support from the local and the Global Learning Centers to set up or help with training projects similar to what we were doing in the pilot. However, we had been cautious about granting any type of support because, at that point, we did not have a solid product to offer. We needed to have the pilot project approved by the steering team before we started any sort of wider rollout, no matter how minor the support might have been. We also wanted to be able to present a full package that was tested and translated and to have installed Learning Center consultants at all sites to ensure the right implementation of the selection, training, and evaluation process of the trainers.

Finally, the day had arrived when these hurdles were crossed, and it was time to begin the wider rollout of the new learning organization. We had built the necessary competencies and tested them by working them out thoroughly in the pilot area. But, the more we knew about the process, the more we had to communicate, and with an organization as large as the LEGO Group, this in itself was a huge undertaking. In addition, rumors had already spread throughout the organization about what the new training process would entail, and these rumors had to be dealt with to make sure there was good understanding of what change was really coming.

Our biggest dilemma starting out was how much control we would need to take or let go of in relation to the upcoming projects. This concerned both local and global-centered projects. As we had spent a lot of time creating, testing, and obtaining approval of the Global LEGO Training Organization, including, among other things, the Standard Competency Framework, Standard Training Method, and the Work Standards (see Figure 8.1),

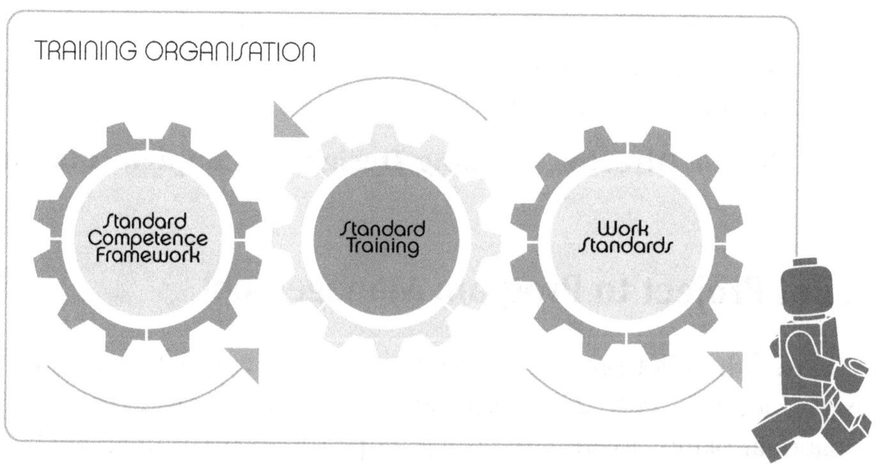

Figure 8.1 Key components to the Global LEGO Training Organization.

we were confident that a solid foundation had been laid, so we decided to let the local projects be run locally as local staff had better insight into the operations that would benefit most from training. It would be a local decision, then, on which areas to start. Once that was decided, we would have them follow the pattern we had set and run the project according to the new organizational scheme for training.

As we pointed out, even as we were running the pilot, there was a pent-up demand for use of the new training methodology, so by the time we pulled the trigger, there were many projects and initiatives that started right out of the blocks. Some of these included the final packaging areas in the Czech plant, where packing was their main work because they did not have any molding production, which was the focus of the pilot project. In Denmark, they immediately went after the decorations and assembly area, where the production of minifigures was the persistent bottleneck to the entire manufacturing process. In Hungary, they went after the warehousing jobs, and in Mexico, they cooperated in a global project with prepackaging work. In all cases, the projects were set up and run in the same pattern as the pilot project, including regular global videoconference meetings with the Learning Center consultants, during which they discussed their work progress and adjusted tools and implementation plans. In other words, they simply repeated the pattern we had successfully experienced with the pilot project and continued expanding the program from one project to the next.

The program steering team wanted all the projects to have a project organization, A3, TIP (Tactical Implementation Plan), and monthly progress reports sent to the committee. This was something we would have to assist them in doing, so we traveled to each of the sites and helped set up local steering groups, the A3s, and the TIPs for their first projects. It was also essential for the projects to have local project leaders as this helps to anchor the projects locally. Here, we tried to have line leaders serve as project leaders because this creates an even stronger anchoring of the projects in the local environment, right on the floor. To gather data to use for evaluating and comparing the progress of each individual project, each month a local steering group meeting was held for each project. The progress reports from these meetings were then sent to us and discussed during a joint global videoconference before we presented them to the overall program steering team.

The program steering team now consisted of the four general managers, each responsible for his or her own site: Denmark, Mexico, Hungary, and the Czech Republic. This meant that we went from three general managers during the pilot project who were in charge of the molding plants to four general managers representing all production sites. This proved extremely valuable to us because this higher level of management generated more attention and dedication from the line leaders. They also needed this increased level of attention as the budget for these projects was significantly higher than the budget for the pilot project, because now we could say just how many trainers were needed and how much time would be used on training, from preparing the training, performing the actual training, to finally evaluating the training.

One of the biggest obstacles to selling the projects was showing the improvement in training time between the old system and the new one that we had developed. We had no problem stating the figures for the new system, but establishing a time perspective for the old one caused us a great amount of trouble. So, we had to devote a lot of energy to convincing management that this was the way to go, and that systematic training will improve the level of knowledge and through this create more stable processes and cheaper global operations. It would be a big investment on their part that would stretch over anywhere from 5 to 10 years, going from a loosely structured knowledge-sharing culture to a systematic and organized knowledge-sharing and implementation organization. This would be an immense cultural change, and it took some effort to push them over the line.

The following **guidelines** were created during program management as a foundation for the Global LEGO Training Organization:

- The Global LEGO Training Organization is an enabler for reaching strategic goals concerning the global footprint in the present and in the future.
- The Global LEGO Training Organization ensures that local sites can sustain and improve their knowledge and skills as needed to deliver on business demands.
- The Global LEGO Training Organization is an enabler to supporting the LEGO Group's step-up strategy: "Speed, Adaptability, Coherence, Flexibility, and Scalability."

In support of these guidelines, the following **objectives** were established:

- Development of a learning organization that can create and maintain robust and transferable processes.

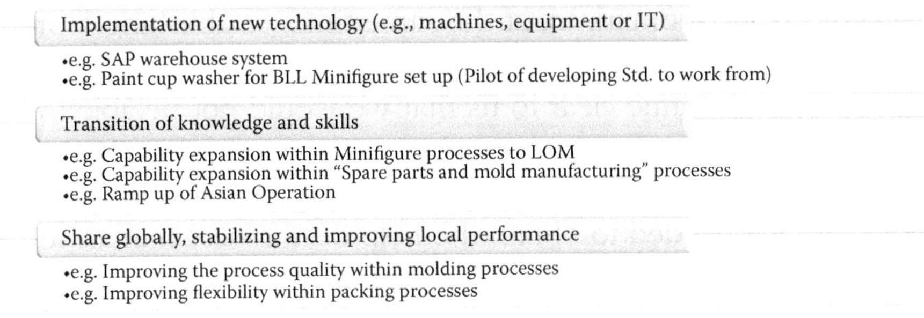

Figure 8.2 Rollout priorities.

- Successful rollout of the tools and processes of the Global LEGO Training Organization and standardized work.
- Effective and high-quality ramp up at the manufacturing sites of the Global LEGO Training Organization and standardized work. This must be done in a prioritized way, supporting first the critical processes in operations. The program steering team conducted these prioritizations for the rollout based on three areas: implementation of new technology, transition of knowledge and skills, and stabilizing and improving local performance through global sharing (see Figure 8.2).

Success criteria for the Global LEGO Training Organization were also made clear:

- Trainers are selected, onboarded, developed, and evaluated.
- Training plans, training materials, and standardized work documentation are developed in English and local languages.
- Leaders are trained and evaluated in their roles in relation to knowledge management.
- It is ensured that local training organizations can execute quality onboarding.
- The local/global training and work standards are developed, implemented, and improved.
- Safety is improved through documentation and training.
- Quality is improved through documentation and training.

All of a sudden, we in the Global Learning Center went from being a secluded small group, who had spent over a year in a development and test phase, to having to manage and participate in a multitude of new projects both locally and globally. Because the sites had quickly established four to six projects each, our work routines also had to change from supporting the local projects in Denmark to, at the same time, establishing a program office that could help steer and coach all of the projects at the various sites. After some time, it became clear to us that we would not be able to both play a strong role in our own local implementation in Denmark and support the other international sites and their local learning centers in their implementations. So, we needed to obtain the help of the local human resources (HR) directors in Denmark to develop a Local Learning Center in Denmark. This would allow our Global Learning Center to focus on the establishment of all international learning centers during the startup phase and better support the sites in future development and within their projects.

Throughout the next year, the work intensified as we prepared for the exit of the Global Learning Center so that the Local Learning Centers could run the training on their own. The focal point for this lay in the fortification of the Local Learning Centers financially and ensuring room for expansion, leaving no room for local management to start taking shortcuts to save money or to move ahead too fast within areas that might not be mature enough for that pace of progress. Managing the speed of implementation was critical in the early stages, so the focus was on stabilization and not going down the road of, for example, linking with the other Lean tools too soon. This task was a hard one, and we did not always succeed.

During this first year, we solidified a structured process for starting up an area that focused on pull from the line organization and cooperation between the Learning Center and the Lean consultants. But, in effect, this was a "one-way street" that was a top-down approach that enforced the pattern that was developed during the pilot project. This created some resistance in the beginning because the new areas did not fully comprehend the learning from the pilot project and did not see the significance of creating stabilized work performance through a well-trained workforce *before* determining standard work. Still, after the first year, we actually ended the oversight of the steering team and let the local sites run the rollouts by themselves. We had created a common approach to starting training in a new area, and the sites were faithfully following the standard project organization, A3, TIP, and progress reports system. Everyone knew their roles, so we could do away with the tight control we had wanted at the beginning.

The more time we spent on systematically working with and developing the training within the organization, the better and more nuanced the terminology became. It was evident to us just how much the organization had embraced it when we were able to consistently speak together using the common language of Job Instruction (JI). For example, when reviewing a problem, everyone knew exactly what was meant if someone asked, "OK, so what is the Key Point here that we are missing?" Even though there were still fights to be fought, the fact that the terminology had been internalized was a quantum leap. We had started a dialogue in which we had raised the bar to a more formalized and effective level.

It would be impossible to generalize all of the difficulties and challenges that were faced in getting the new training organization firmly embedded in each production facility. However, in Chapter 9 we present several individual case studies of what key players experienced in this endeavor. These will give a sense of what it took and the great effort that was made.

When we look back at the 3 years the Global Learning Center used on developing, testing, and supporting the global job training organization and the Local Learning Centers and then fast forward to today, it is amazing to see how it has grown in size. Today, there are more than 300 job trainers developing material, training, and evaluating more than 3,000 employees systematically and with real precision. Furthermore, there are Learning Center consultants connected to each plant, all in charge of the selection, the training, and the evaluation of the job trainers, all done following the exact same concepts. This process continues being developed and undergoing further rollout. This is where the local and global long-term stabilization and standardization starts, through the learning organization.

Job Relations Implementation

The implementation of the Job Relations (JR) module of Training Within Industry (TWI) started in the LEGO Group 3 years after the JI implementation, but with a different approach. We had been talking about adding JR to our global training initiative for some time and had already started discussions with our HR team about its effectiveness in supplementing the JI program. In fact, Bob Wrona from the TWI Institute had initially recommended that we introduce JR right from the start as a needed addition for success with our JI effort. At that time, we told him emphatically that we felt the LEGO culture was covering this aspect of the leadership cycle well, and we wanted to concentrate solely on the JI module, which was in our target sights.

We found out fairly quickly, though, the wisdom of his approach when we realized that our supervision team did indeed need practical leadership and relationship-building skills to have workers follow their good instructions. This need was brought to light when we began the JI and found that what held us back from achieving true standard work was not necessarily just our instruction skill but also our overall leadership abilities. As we learn in JR, if our people are not "following" our instructions, then we are not "leading." Nevertheless, the production sites wanted to "keep their eye on the ball" and not become distracted from learning and implementing the JI method, so we waited this long before introducing JR.

The master trainers and first-line production supervisors from our existing training organization were the initial target group for using JR. We felt that these were the groups that directly influenced the workers and their willingness to cooperate in the creation of good work standards, not to mention the development of an overall positive working environment. We decided going forward also to include first-line leaders in this effort. More often than not, the first-line leaders are people who have progressed from operator to mold setter to technician and, for the most part, have not had formal leadership training, yet they possess vast practical and relevant production experience. Including these first-line leaders in the JR development program has been overwhelmingly positive as these employees are eager to learn, and they value these management and leadership courses, which they have never had.

A good example with such first-line leaders occurred in the molding factory in Billund, Denmark. There was a leader of two different weekend shifts with some 42 employees reporting to him; he is educated as a toolmaker and has been working with the LEGO Group for 7 years, having started in production as an operator. Eventually, he was promoted to a technician and finally worked several years as a setter. In that position, he was selected and educated to become a TWI job trainer and now is in a prominent position as a production leader. He recognizes the benefit of having worked in the technical and mechanical aspects of production as he gained the needed knowledge and insights to be able to understand and answer inquiries and address issues arising from his employees.

This leader was part of the first group in Billund who went through the 10-hour JR class. He was excited to join the class, but here were his first impressions of JR:

> To be honest, in the beginning it was a little difficult for me to see what I should use it for. I did not see that much new and different compared with what I did already. After the training, I found out that grabbing the card from the pocket is maybe not my strongest side as I do not find this natural. When I was a TWI trainer, it felt very natural for me to hold the Job Instruction card

> in my hand while following the method during a training session. But with Job Relations, it is different.

The leader told us about a case that occurred in production for which he was able to apply the JR training and this changed his opinion of the value of the method. There had been an episode with an employee who had made an error that resulted in a major quality defect. He said that he had no doubt that it was the employee who was responsible for making the mistake, and because there had been some issues with this employee previously, he had decided to give the person a warning. He felt that this was necessary not only because an infraction had occurred but also more importantly because it influenced the professional pride and respect of the team overall. But, because of his JR training, he realized now that he did not have all the information he needed to make a decision. He explained:

> I missed talking to the employee himself. Even though I am very detail oriented I failed to ensure I *had the whole story* before concluding on the necessary actions. Luckily I was told by the job stewards that they had heard another version of what had happened and we arranged a meeting with the employee. After this meeting, I had all the nuances of the story and the meeting provided me with the full picture of the problem, and it turned out that it was not only this employee who was the reason for the quality problems. What is more, there was no reason for giving the employee a warning. I learned a lot from this case as I was very close to *jumping to a conclusion* without having the whole story of the problem. I must admit that this gave me a lot of respect for the process and not taking hurried decisions even though sometimes it is a challenge to communicate with the relevant people across different shifts, etc. I think that the practical difficulties pushed me a little in this situation and I was very close to taking the wrong decision. I am very happy that I got the process stopped in time and I avoided giving this employee a warning on a wrong basis. I have now really understood the importance of *hearing all opinions and feelings.* (italics added)

In addition to the 10-hour JR class, what we like to call "Learning-by-Sharing" sessions have also been organized. The purpose of these sessions is to help the JR practitioners learn from each other based on their experiences using the method. These sessions also provide needed coaching, mentoring, and networking with the implementation of JR. In the JI implementation process, all job trainers were carefully selected and then closely coached and provided with several follow-up activities. The JR process, on the other hand, has been delivered as an open opportunity for first-line leaders with follow-up and coaching provided in the Learning-by-Sharing sessions. This integration of the knowledge, methods, and skills of the JR program can eventually create constructive progress in the production environment.

The Learning-by-Sharing sessions include 8 to 10 employees, and the cases presented are reviewed step by step with the facilitator following the standard JR practice format, going through the four-step method card with participants contributing ideas and input to learn from the specific story being covered. The facilitator is one of the Learning Center consultants, and his or her task is to ensure a secure, stable, and trusting environment that allows each participant to confidently bring forward and present stories, just like in the 10-hour class. This is important in dealing with personnel issues. This confidentiality is also why we separate the classes for first-line leaders from the classes for job trainers so they will be willing to talk about their problems with no fear of embarrassment or judgment.

The same leader mentioned previously was also part of a Learning-by-Sharing session with other first-line leaders. It is revealing to hear in his own words how he is responding to the JR method, and you can see (in the words we have italicized in his quotations) the parts of the JR method he had internalized in his dealings with his employees:

> We are a group of leaders meeting to share experiences around the use of Job Relations, and this is very beneficial for me. The 10-hour class is naturally a precondition for learning the method, but the learning sessions are giving me much more. We use our own examples from the shop floor and get many more insights when we are in a group as opposed to just doing it on our own. Especially when we talk about *which rules and customs apply*—this is one of the points where it is very useful to get some sparring from my colleagues. It is a process where we go through the method step-by-step and more information and more details are always coming up from the discussion. What becomes very visible is that it is about ensuring a holistic perspective—what is the *best solution for the individual, production and the group.*

We meet every second month where we have a formalized session organized by the Learning Center. So we practice the method in the learning sessions. In addition to this, we as leaders use each other to a great extent, and it is very beneficial to share and support each other in the problems we have to deal with.

One of my primary reflections has been the importance of and what it takes to *build relations with my employees*. I try to *create good relations with employees* by talking to them. Every day I walk around on each shift. I am a very fact-oriented person and there are many requirements as to safety, quality and delivery. With 42 employees in 12 hours, there is not much time to focus on building relations through small talk. In a dialogue with employees, there is a lot of focus on the facts related to our deliverables. So I have learned that it is very important to be aware and attentive to what is going on in addition to this. And talk about our deliverables issues might be pointed out indirectly. If, for instance, issues are coming up around collaboration, when people cannot work together well it is important that I focus on that as well.

One of the foundations I also focus on is to *tell my employees about changes that will affect them*. We have many examples where there are negative reactions if they are not informed or if they are informed too late. Half a year ago, it was decided to change the system from each employee having their own personal tool wagon in the production modules to having shared and common fixed tool columns instead. These are everybody's daily working tools so this change has really been difficult. Because our machines are getting bigger and bigger, we have less and less space and it was decided that there should be fixed tool columns in all modules. No one likes the solution and I have used a lot of time *explaining and talking about it*. I can see the need and importance of *involving more people in this change and earlier*, as this is very fundamental to their daily work. I try to be proactive when changes are coming. On this change we could have been more thorough and proactive in our approach. (Italics added)

As we can see from this case, the first step in the JR implementation has been completed at all production sites. The next step is to implement the program for the full job trainer organization.

Rollout in Asia

The LEGO Group will soon be taking the next decisive step toward achieving its ambitious growth target in Asia, and the global learning organization now in place will play a significant role in ensuring its success. A new factory will be constructed in the city of Jiaxing in the Shanghai region and will house molding, processing, and packing departments. This factory will produce 70–80% of the LEGO products required for the growing Asian market. The factory layout will be built on a "copy-paste" approach, as formulated from our work with LEGO specialists worldwide, to duplicate the success of our plants in Denmark, the Czech Republic, Hungary, and Mexico.

The plant's initial machines are expected to be installed by 2015, at which time the workforce will number 200–400 employees. Once fully functioning, the factory will employ approximately 2,000 employees. Although Asia is an immense region offering unlimited possibilities, the city of Jiaxing was selected as the site of the new factory because it best matches the objectives of the LEGO Group's vision. The area is a good representation of the total Asian market, as well as being a "green" city offering pleasant surroundings and fine housing for LEGO employees. There are no plans for employees to live in dormitories at the plant, which is typical of factories in China, but rather they will be provided regular housing accommodations. In addition, the plant will be close to the LEGO regional distribution center for Asia in Shanghai, which will significantly reduce the lead time to Asian customers.

The establishment of the new plant in Jiaxing will not affect any other LEGO plants in other countries or current suppliers in Asia but will rather provide the opportunity to meet the growing demand for LEGO products already experienced in the region. Because of this, opening the plant is not a cost-cutting exercise with the only focus on manual operations done by cheap labor. The company will install the identical machinery and operations that are currently used in LEGO plants in other countries, including Denmark.

Furthermore, the new facility will have the same requirements as its sister plants worldwide, including a high level of safety and quality. Conditions in China are different from other parts of the world where the LEGO Group operates. China's economic environment can be both complicated and difficult to navigate. Therefore, it will be crucial to attract and retain quality employees while providing competitive salaries, ideal working conditions, and a positive workplace culture.

As we have stressed throughout this book, the global objective of the company is to *work strategically and proactively to ensure that we have the right people at the right location at the right time and that costs are in alignment with business acceleration or deceleration.* So, when launching a new plant, based on many years of experience starting up production in other countries, we know that the training and development strategy must clarify just what the educational level of people in the area needs to be so that we can attract the right people and competencies and develop them as needed to meet the requirements of the company's high-quality production standards. The Asian growth strategy of the LEGO Group requires a vast expansion of production capabilities, so it is important to keep a strong focus on how the expansion is approached and initiated. In this case, we now aim to create in each location an on-site educational development infrastructure to meet the required educational levels in each specialty and profession.

The ability to train people in complex processes that were traditionally kept in the Denmark facility is an example of how this localization of education works. LEGO Operations has thorough expertise and experience in building low- and medium-complexity processes on a global level. However, when production sites are scaling up as quickly and extensively as will be the case in China, the high-complexity processes have to be built alongside the more basic ones to ensure that the sites are self-sustainable. Based on our experience with the fast startup of the Mexico plant, it became clear that having foundational educational programs for toolmakers and plastic makers was a precondition for self-sustainability in relation to, for example, spare-part manufacturing and ramping up of molds. Without the foundational capabilities within the mold and molding processes, the capability platform would be fragile and incomplete.

This high-skills training, then, must take place in the country in which the work is being done. In Denmark, Germany, Austria, and Switzerland, there is a long tradition of educating toolmakers, the people who create the steel molds that are used to produce the plastic LEGO bricks, which is a high-precision set of skills that takes years to learn to do. This "best-practice" toolmaker education requires 4 years of training to fully understand common practices and gain adequate experience. To become a competent LEGO toolmaker, this process takes at least 7 years. Therefore, to re-create this process in the local communities where our overseas plants are located would require an ongoing, close collaboration between these local plants and vocational technical schools in their areas.

To ensure the success of this educational process for apprentices, the critical part is managing the combination and coordination of hands-on practice during the school periods with LEGO practice periods in the plant. In other words, the trainees' hands-on practice at school and at the LEGO plant must be well coordinated and reinforcing—what they practice at school reinforces what they practice at the plant and vice versa. Furthermore, we must also ensure that the theoretical platform is maintained, and the same skills are continuously practiced. In addition, it is important that the apprentices have more experienced colleagues in the plant who can act as mentors for them during their practical periods in the company where they work with LEGO-specific tasks.

The plastic maker education is also done through collaboration with local schools. At this time, there is no best practice for plastic maker education; therefore, the Billund plant is providing the necessary training, guidance, assessment, and school networking to develop the required knowledge and skills used within the injection molding processes. The complete training requires 4 to 6 years, also alternating between the local school and the local plant. During this time, there is continual exchange between local production and LEGO molding in Billund, Denmark, where coaches are available for additional training.

Insight from the Mexico Experience

When our factory in Mexico was starting up in 2010, the Danish consulate assisted in finding relevant vocational technical schools for collaboration with the LEGO Group. We wanted to attract the best-qualified people for the plant. At that time, we were not yet aware of the importance of finding the right schools with the required educational level. Based on our 80 years of molding experience in Denmark, we had assumed that the craftsmanship needed to become a toolmaker was a normal ability that was being developed all over the world. However, the more complex processes required in the new LEGO plant highlighted the difficulties of this craftsmanship ability, and within 2 years, it became apparent that the training of the workforce was insufficient.

We had, at first, tried to collaborate with local schools that were close to the plant, but it was not easy finding vocational schools that had the necessary resources in relation to machines, equipment, and teachers. Most of the schools were theoretical and less practical in their approach. In many of them, the apprentices learned the theory of toolmaking, but the "handling" of the machines was only done through observation, not by hands-on training. With this kind of program, the workers were not prepared to perform the complex processes required as production at the plant grew. With these problems, we initiated a more thorough search of vocational schools with the support, again, of the Danish Consulate in Mexico, extending the geographical area to include more schools. Even with a larger scope, it was still difficult to find a technical school with our required level of educational programs.

During this time, we discovered that one of our German machine suppliers was delivering machines to a German technical school in the Mexico City area, some distance away from our plant in Monterrey. This school was then reviewed and evaluated, and we discovered that it had an exceptional educational program for the basic parts of toolmaker education. An agreement was established, and our Mexican toolmaker apprentices are now attending the program at this school. Of course, this is done in coordinating the practical training they receive at the plant with their program in Mexico City. The final training is done through a specific curriculum designed at our Denmark plant.

Drawing on our experience in Mexico, we initiated a thorough search for the best technical schools near the site of the new factory in China, and through suppliers, other multinational companies, and our network of business contacts, we found nine vocational schools to be reviewed and assessed. We visited them in October 2013 and found that they were of exceptionally high quality, providing just the kind of craftsmanship training we will require. We understand the lead times necessary to develop the specialists, and we know the importance of having workers ready to work and others in the process of development. Again, the two core areas of plastic making and toolmaking are the priorities. The schools will be evaluated on their approach to teaching the right combination of theory and practice, the availability of relevant machines, teachers' backgrounds, and how the students learn. Once the appropriate schools are selected, agreements will be established, and educational activities will be initiated. As other competitive companies are also attracting students as well, we know that once we enter the schools and meet the prospective apprentices, it will be imperative for us to highlight the benefits of working for the LEGO Group and encourage the best among them to join us at our new plant in China.

Training Approach

The educational level is crucial because we must develop a workforce with standardized training so students learn the skills required for all the positions required in production. Launching the plant in Asia will be a different experience for us considering the existing training infrastructure already established in Asia. Integrating these local facilities into our newly established Global LEGO Training Organization will be a challenge. Stephen Burke described the coming task in China:

> We now have a standardized way of building up and transferring knowledge and skills to China, and this will be a great advantage for us to build up the factory in an efficient way. There will be cultural challenges which we will be able to focus more on. It could be a challenge whether, for instance, our trainers will have the needed authority in the organization.

However, with our new standardized training approach, we will be in an advantageous position for starting up the new plant. By reviewing the number of machines needed and the timing and number of employees required, we will have the opportunity to prove that the training organization we have established will work. When creating the business plan for the new China plant, we used input from all of our functional area master trainers (FAMTs) regarding each of their own site's costs and timing of their training. The master trainers from Mexico, Hungary, and Billund all provided feedback regarding what it takes to provide the best-practice training plans for key positions, such as operators, material handlers, color changers, and mold setters.

We also looked at all sites to determine where we would be able to best perform the startup training for our new Chinese colleagues based on competencies, machine setup, and other factors. First, we did not want to assume, as we would have in years past, that this initial training should or would be conducted without question at the plant in Denmark. Reviewing each site's results, we considered doing some of the training in Mexico, which has a physical setup similar to the future site in China. As the Hungarian factory was in the process of expansion and relocation, it could not provide realistic training situations and examples, yet the site could still prepare training materials and assist with business results. Therefore, we would train the new Chinese employees using the facilities as well as the expertise from *all* of our molding plants.

Using this input, we needed to arrange our best-practice training plans. To begin, we chose to have a face-to-face meeting with all the master trainers from all sites. Through a 2-day workshop, we hoped to formulate the best-practice training plans. We divided the FAMTs into groups by their functional area, including molding operations, mold maintenance, and tool shop. Each group worked with a Learning Center consultant, and they all prepared a presentation outlining their local training program and activities. Using the existing states of their competency overview, skills matrix, and training policies and dialogues, the best-practice procedures were defined for each position. In addition, a detailed plan for when and where each phase of the training should occur was determined, as well as which of the existing global job trainers would train the new Chinese colleagues.

Based on the staffing projections for China, the first step in recruitment was to hire a core team of workers with knowledge and experience in the basic positions required for production. The core team included technical writers as well, who also function as translators, and a Learning Center consultant. Both of these positions monitor the training that the core team receives to acquire the necessary knowledge of the business. From within the core team, we looked for people who had the potential to be good trainers based on the defined trainer competencies we outlined in Section III of this book. A trainer selection process was then initiated to find the new global and local trainers for China, following our global learning structure, and these people actually conduct the training for the next wave of employees. In this way, the Global LEGO Training and Learning Organization serves as the backbone for developing the first generation of workers at the new China plant, and because the content of the jobs has already been defined and documented in the other plants using the TWI format, the operations will start up quickly and smoothly based on the TWI skills already internalized by the organization.

Using our existing global training organization and the standardized training system described in this book, we are focusing on providing professional training for the new workforce in China as well as creating a sustainable training organization at the local level.

Chapter 9

In Their Own Words: Case Studies from the Implementation

Introduction

We have endeavored to explain throughout this book the details of the process the LEGO Group went through to create and sustain a system that would indeed help it to become a true global learning organization. It was an incredible journey for all involved. But every human situation is unique and no one can anticipate what challenges will be faced when implementing organizational change. With that in mind, we would like to finish by highlighting some of the more personal endeavors and efforts made by just a few of the key people who made this all possible. We could never cover nearly all of them, but a small representation of what it took on an individual level to ensure that the overall project crossed the finish line is both enlightening and inspiring.

Ericka Hernandez: New to the Learning Center in Mexico

Ericka Hernandez was brought into the LEGO plant in Monterrey, Mexico, as a Learning Center consultant after the plant's first member of the new training organization, who had been trained with the initial group in Training Within Industry (TWI), suddenly had to leave the company. So, she got off to a late start, having missed all the initial orientation and training meetings. As a key member of the worldwide training team, her main responsibility at the plant is to implement standardized work and to establish and support a local training organization to ensure that all the necessary skills are in place in the workforce. Erika explained her role as follows:

> In Mexico it is a challenge to work on developing the workforce. There is a tendency to view this as simply a formal process that only takes place on paper—in practice, it is difficult to get the resources to accomplish anything concrete. The LEGO Group takes a different approach and there is a lot of focus on human development and training of the workforce. It is still a challenge, though, to change the mind-set of the leaders and it is very supportive to have clear direction from top management that skills, people, and training are all important elements.
>
> The Mexican employees also see the benefit of this. In interviews, employees have responded that this is the first time they have received such extensive training in a structured way. People also say that this is the first time they have experienced trainers and leaders making sure that they learn to do the jobs in the correct way. In the beginning, they felt a bit under pressure with the method having to remember the Important Steps, the Key Points, and all the Reasons for the Key Points. But in the end, they liked the process because they know that it is important. That is why they are participating.
>
> I started taking the 10-hour class in Job Instruction on my very first day with the company. I joined both the group of trainers participating in a morning session as well as the group of trainers participating in an afternoon session. The reason for this was that in the very near future I was the one who was supposed to train this class. The training was conducted by an external TWI trainer from El Paso, Texas. This methodology was very interesting and new to me and taking the class was a very good experience. Although it was different from what I was used to, I liked it. So I have been doing TWI since my very first day at the LEGO Group.

The initial group of Learning Center consultants had already taken the TWI trainer development program, not to mention the 10-hour class given several months before that, so that they could begin delivering the Job Instruction classes locally at each facility in their local languages.

But, because the Mexico facility's consultant was no longer with the company, they were forced to bring in an external trainer from El Paso to conduct the training. This was the initial week of the training delivery at the plant, and Ericka was hired just in time to participate in the classes. That meant she did not have a head start on the rest of the organization she was expected to lead in the training effort.

> So after that initial week of Job Instruction training, I participated in a global workshop with my new colleagues from all over the world—the other participants of the workshop already knew each other based on close collaboration over the previous 4–5 months. It was a tough process to learn about the LEGO Group and try to understand the project at the same time. The language was sometimes a challenge for me but I wanted to contribute and give my input. I was not sure what I needed to do but I knew that my manager had a lot of expectations of me. It was a very hard week and I must admit at the end of the week I was terribly tired.

As the new Mexican Learning Center representative, Ericka was participating in the global alignment process and coordination of the pilot implementation of the Job Instruction training. After the 1-week global workshop, the process of starting to develop the documentation was initiated. This gave Ericka a great opportunity to be on the shop floor extensively and to learn the jobs and processes.

> Now I know how important it is to collaborate and to share experiences and best practices. The key is to share to make it better. By opening your mind to hear new ideas to improve local processes and to adjust to the local needs, that is the great benefit of this work and it is a good thing for the company.
>
> This was a beneficial learning process for me. I really consider it a privilege and opportunity as an HR [human resources] staff person. From the beginning I was on the shop floor, getting to know the people and how they are doing the jobs. My task was to support them as trainers but, at the same time, I learned a lot. We were on the shop floor all day guiding the trainers through the JI [Job Instruction] methodology and, together with one of the Global Job Trainers, we were very focused on both breaking down the jobs and practicing the training method. It was very

> good to collaborate with the Global Job Trainer to secure the correct way of implementing the method. I was very focused on how to make a good job breakdown—a skill I like to apply and help trainers develop—and also to maintain the discipline of the methodology. Both factors are really important. They can learn to follow the method and, at the same time, they are testing the job breakdowns.
>
> There were many conflicts and I think a critical point was how to keep the team together and how to handle the differences in mind-sets, ways of doing the job, etc. So as a facilitator, I focus on ensuring a good process around this with the trainers to ensure consistency, sharing of input, etc. In the process, I ask one of the trainers to do the job while a couple of the others are taking notes and observing. I follow the standard procedure of breaking down the job: Has the job advanced?, What did you do?, etc. Then we try out the first version of the job breakdown by training it, adjusting it, and training again, adjusting it again. In that way, the trainers learn the method.

After a month, Ericka went to San Diego, California, to attend a Job Instruction 40-hour train-the-trainer session to learn how to conduct the 10-hour Job Instruction class for Mexican trainers. After that week, Ericka began conducting the training. In a period of 3–4 months, she also received the DiSC certification and a train-the-trainer session in the trainer selection process as well as the Training and Learning Skills Course. So, within that short period of time, she was up and running on the basic skills and knowledge of the Learning Center consultant job. It was an intensive introduction period.

> It was a hard process and the company was new. And I had to deliver classes right after the trainer development sessions where results were expected immediately. All the while, I was also trying to learn about the company. Normally when you start in a new company, somebody will be there to teach you and guide you but here I was learning and executing at the same time. Now that I look back on it, it was a challenge. At the same time I think it was the best way to learn it. I can see that when we have new people in the Learning Center they get a smoother onboarding process. It is easier for them to ask me how to do it and get advice.

> But the only way, and the best way, to learn it is by doing it and trying it out yourself.

After a period of 3 months, Ericka received her permanent contract with the LEGO plant in Mexico. She explained how she overcame the intensive startup process in her job:

> I was very excited about the new job, I was very committed to it, and I wanted to prove that I was the right person. My previous job was very different, so I was very eager to start working with training again and saw a great opportunity for my personal development as well. I had a lot of pressure from my manager over performance, but at the same time I had room to develop the tasks myself and that inspired me to make the effort. Furthermore, I wanted to be part of the Global Learning Center team as there was a good energy there and I was very motivated to be part of this group. The challenge of making a difference and wanting to prove that training is important and needs to be in focus created my motivation. Sometimes there is a perception that training is just a waste of time and resources. But in the LEGO Group you have the opportunity to show that training is really important. I found that mind-set here and really like it.

Fruzsina Veress: Leadership and Implementation in Hungary

This case involves the experiences and challenges of Fruzsina Veress, Learning Center consultant at the Nyíregyháza, Hungary, plant, in her implementation of the Global LEGO Training Organization there and the role the human resources (HR) and Lean departments played in the process. She was involved in the TWI project from the very beginning of the implementation phase from a development perspective. She participated in the global development and local implementation of selecting and training internal job trainers, and she played an important role in launching both the pilot and rollout phases of the standardization project at the Hungary plant. Her main objectives were to include stakeholder management in the plant's leadership team, to set up the best fit and function for the training organizations in different production areas, and to make sure that the people they dedicated to training tasks were equipped with a proper trainer skill set. Moreover, she wanted to be sure they had continuous HR support while engaging in training and mentoring activities.

After the pilot project was implemented in the molding department and the global workshop was completed, the successes and difficulties of the process were evaluated. The experience and knowledge learned helped to develop a plan that best matched the organizational needs. She wanted to sell the concept to the rest of the leadership team to start up the rollout in other parts of the factory. According to Fruzsina:

> The idea was to make this campaign step by step, so that we don't flood the organization with an improvement which is not totally tested and aligned. Although there was a need from the production directors to create standardized work and a training organization in their areas as well, their imagination about it was a bit simplistic. Therefore, first we had to ensure their full understanding of the concept. Then we gradually built up the framework in all key departments, making sure that it was adjusted to the area features and needs.

The need for standardization was quite obvious in Hungary as the shop floor processes were neither aligned nor uniform. The procedures had been either created centrally in Denmark and then introduced locally with a low chance of adjustment to local characteristics, or they were developed purely based on local knowledge and experience, leading to large variation in processes, quality, timing, and safety. Thus, the standardization of production was necessary.

> That's how we could make our procedures measurable, controllable and improvable. TWI provided a good tool to make all this a reality. However, we did not really have any previous experience with this, hence it was not easy to estimate the resources we needed to put into this project. That was one of the difficulties we faced when trying to sell the concept of the training organization to management. Also, at first it was challenging to implement the discipline around the application of the TWI methods since the approach was a bit strange and unnatural compared to Hungarian cultural characteristics. We had to see and show some advantages to be able to convince production people to accept and use the methodology.

In Hungary, the top management had a clear idea of what they would like to see as an outcome of the project (standard processes, efficient training of new and old employees, more effective production); however, they underestimated the resources that they would first need to invest to obtain the benefits. Therefore, some research and analysis were conducted to show just how much needed to be put into the program to reach the desired results. These studies were not enough, though, to totally persuade them to change to a new training system. They were engaged to the idea theoretically, but not totally committed to the project from a material point of view. It was even more difficult to convince middle management as these leaders were not really aware of the whole concept of this improvement. Thus, there was resistance. Fruzsina explained:

> The only way to get to their heart and mind was to get them involved and give them some responsibility in setting the goals in connection with standardization. As soon as they could understand the overall purpose and also their role in the implementation, we could more easily cooperate with them. It took quite a long time to create a good image around the standardization project in the organization, but with endurance, systematic arguing, and involvement we succeeded.

The level of acceptance started to increase quite dramatically once the first results from the pilot area were seen. That was the real selling point in the history of the project.

At the beginning of the implementation, there were some conflicts in terms of who would be leading the implementation: Lean, HR, or the production leaders? The key here was to find a good balance and realize that everyone can offer something. However, the real need and goal setting should come from the leader who provides the resources, in other words, the production leader. The knowledge, experience, and tools would come from the supporting areas. We do not want the cart to be pulling the donkey, so it is imperative that this balance be created and maintained each time when implementing in a new area.

One issue that was exposed in this dilemma was that the Lean organization was not involved in the project from the beginning. They could have played a vital role in the implementation by providing and developing the tools to be used throughout the standardization process. Instead, there was an initial lack of interest, attention, and energy on their part.

It became obvious that they needed to play an active and important role, involving them as a full-fledged partner in addition to the HR agents. Once we began teaching the organization how to use an A3 or a Standardized Work Chart, the Lean group started to contribute to the project with full resource investment and engagement. Fruzsina's observations were as follows:

> As I see it, one tipping point was when the supportive functions (HR, Lean) realized that it is basically not in their interest to push the implementation through the leaders or their production areas, but actually it should be the other way around. The assignments should come directly from the production leader. Of course, this need must be developed wisely so the change agents can inspire these production leaders toward standardization, but the latter should articulate his or her explicit need toward improving the quality of both job skills training and the way of working. Standard work is a toolkit which we can offer to reach these goals and adjust based on the actual needs. So I think the attitude is key here—that HR or Lean should not direct or lead, but instead guide or facilitate the leader's commitment or mindset while making sure that targets are set and tools are available.

The Hungary plant was constructing a new factory, and this was a motivating factor for boosting standardized work there. Easy and controllable processes are quick wins to attract employees to a new facility. With the planned introduction of autonomous work groups in which workers can plan and check their own work activities and have a direct impact on improving procedures, standard work will be a real success story in the life of the LEGO plant in Nyíregyháza, Hungary.

Pavel Kroupa: Key Roles and Collaboration in TWI Implementation in the Czech Republic

This case is about how the standard work and TWI activities were rolled out in our production site in the Czech Republic. It shows how Pavel Kroupa, Learning Center and Lean consultant, ensured collaboration and stakeholder management at different levels in the organization. Since 2010, when it was decided to implement standard work and TWI at the Czech plant, the program has been rolled out in the areas of decoration, assembly, prepack, final pack, warehouse, quality, and maintenance—processes covering the full production of what they do there. Some of the areas are fully up and running with standard work, and some have just started. So far, 280 trainers and leaders have been trained in Job Instruction and 40 leaders have been trained in Job Relations.

Even though the initial global TWI pilot project was in the molding area, and the Czech Republic plant does not have any molding activities, Pavel joined the pilot to be part of the Global Learning Center collaboration and fully participated in the development of the tools and standards that became the Global LEGO Training Organization. As he explained it:

> I remember clearly the first global workshop in Mexico when my manager asked me whether I was ready to go. We had a clear agreement—if it was not interesting I was ready to pull the plug and say "no" to participating in this project. So to be honest, I did not have any expectations for this. But it was really an eye-opener for me as I have a very long and solid Lean experience and was surprised I had not heard about TWI before. The TWI method is easy, it is understandable for everyone, it is common sense, so it is really an absolutely fantastic way to train people. I went back from the workshop full of energy, enthusiastic and eager to work with this project and, even though the pilot was in molding, I was not going to wait for our implementation. So after the workshop I made a 360 degree change—this was what I wanted to do. I had a meeting with my manager and we started the communication and sessions with top management to create our plant's participation in the project.

Collaboration with Top Management

In the beginning, it was a challenge to convince top management about the potential of standard work and TWI. They did not know much about it. Their reactions were based on the resource requirements and, in general, it was a different way of doing standard work so there was also a resistance to the change.

> The way I handled this was by giving an "appetizer" of Job Instruction which was a short management presentation where the top management team and other key stakeholders were invited.

> I presented the method and made a demonstration of the Fire Underwriters' Knot and I chose our General Manager for the correct demonstration of the method as done in the JI class. I also shared some of the benefits and results of TWI based on material from the TWI Institute. On another occasion, I had the chance to make a simple demonstration of Job Instruction for our Executive Vice President for Operations when he was visiting our plant.
>
> Another factor which influenced the change process was the fact that this was a global implementation and the General Managers from all the production sites were talking about it in the program steering team. This created an awareness and sharing of how far each plant was in the implementation and I think a little competition was initiated. This influenced the project and suddenly there was pressure to do this implementation.

Another activity that was initiated was one-on-one sessions (what the Japanese call *nemawashi*) between the Learning Center consultant and relevant leadership team members. Here, the program and its potential were introduced and what would be required in terms of people and time. The leaders who were selected for the first initiatives were chosen not only based on their functional areas but also based on the trust that already existed between the leader and the Learning Center consultant. So, the collaboration was built on an already-existing strong relationship. The objective of the *nemawashi* activities was to obtain agreement with the leaders ahead of time that they would support the TWI implementation and allocate resources to its implementation. Furthermore, they agreed to give it priority, and the activity was added to their Tactical Implementation Plan (TIP) for their area. In that way, the commitment from the leaders was secured. Immediate follow-up activities were introduced through weekly huddle meetings using the training boards with follow-up on the TIP, obstacles, and so on with the management team and the master trainers who presented the progress for the planned activities.

Collaboration with Production Management

The initiatives that were planned for production management included the 10-hour Job Instruction class, which was conducted for relevant area managers and shift leaders. Even though these managers would not actually be training jobs, this gave them an understanding of how the method works as it is hard to understand the details of the method and the logic behind it if you have not participated in the class. In addition to this, the managers involved made a draft of their Key Performance Indicators (KPIs) in collaboration with the Learning Center consultant, which was a hard process as it had just been started, and they were new indicators to measure. An example of an indicator would be training time in prepack, where the target was to reduce training time from 12 weeks to 6 weeks. Furthermore, practical problem solving (PPS) was also part of this effort as it contained the elements "lack of standards" or "not following standards." This root cause originally involved 80% of all problems, and the goal was to reduce it to 50%. Both of these targets were reached.

> Before, when a new operator started, he or she was told to go to a more experienced operator for training. But as the more experienced operator was busy running the line, the new operator was often asked to sit down and do things like sort rejects. After several days doing this kind of work, the new operator was confused and tired of the job and very often quit within the first week as they were not learning anything new and did not like the job. Now we have a formalized training organization and the trainers are dedicated to specific new employees. We also have training plans for each specific job. We know exactly what the new operator will start training on Monday morning at 8 am and what training is planned from 9–10 am. They have clear plans for each week of training and the onboarding process for each job and the different job levels. So within the specific levels of the jobs, we know exactly what jobs people are able to perform and we have clear directions and a plan for our high season when we know that we will need an experienced operator and a temporary operator working together to cover the line. This makes it possible for us to onboard many new people in a short time.
>
> After a period of time I started challenging the leaders: How do you actually know if the people are following the standard processes? Before TWI, the follow-up on work processes was based more on feelings than on standards. Now we track the number of successful process confirmations made by management in their areas thus making the follow-up based on standards and not on feelings. Furthermore, we measure on the standard work maturity of the areas based on specific parameters like documentation started, basic knowledge defined, skills matrix defined, selection of

trainers made, evaluation of trainers done, etc. In this way, we have a clear picture of how far each area is in the process.

For 2013 we have a Performance Management Program for all hourly paid people, and 40–60% of the bonus in the program is related to adherence to standards. Also, the production management team has standard work as a part of their performance program as well.

Ole Therkelsen: Mold Manufacturing in Mexico

This case study is about the TWI experiences from senior director Ole Therkelsen and how the LEGO Group is working on developing capabilities within spare parts and mold manufacturing processes in its Mexico plant. Ole has been working with the company for 28 years and heads the department of prototype and mold manufacturing with 80 technical specialists located in Denmark. They are responsible for manufacturing novelty molds for production and ensuring that the final LEGO product meets the designers' needs in element design and geometry. This requires a high degree of craftsmanship in refining the molds, the core competency of the area, with advanced milling processes and wire cutting and spark erosion technologies that meet high-precision tolerances down to the micron level.

Ole's team was tasked with a project to support capability expansion in the Mexican molding plant to equal the size and breadth of the Danish molding setup. To make the plant self-sustainable, we had to develop molding competencies and spare part manufacturing at the plant. Ultimately, the goal of the project was to have the Mexico plant be able to produce simple molds and more complex spare parts for production.

As the global technology owner of processes within CNC (computer numerical control) milling, Ole's role was to support the Mexican organization in building their ability to handle these technologies. To help them accomplish that, an initial pilot project was started in his area to implement the TWI methodology. Before the pilot project was started, an introduction meeting was arranged with Ole's manager and the Learning Center consultant. At the meeting, the background for the project was clarified and the Learning Center program introduced, including trainer roles, resources required for the project, and so on.

The first time I heard about TWI was at a meeting around three years ago. I must admit I got a little scared when I heard about

> it and what it would require to work with this new method. I was asked to join a couple of meetings regarding the new initiative and, after being challenged, I finally agreed to join in and allocated resources to the pilot project. The objective of the project was to build up capabilities in our factory in Mexico using TWI as a methodology. At first, I was only committed to trying it out and delivering on the pilot project for a period of 6 months.
>
> During this same period of time, I was on a business trip with a good colleague of mine. He told me that he had joined a lunch session with a TWI specialist where the TWI method was demonstrated. Furthermore, during this session he had been taught a job as a trainee and he could very clearly and precisely explain the job he was trained in. He had been really impressed and this influenced me.

When a new area is starting with TWI, we emphasize the importance of the leaders of that group taking the 10-hour Job Instruction class to ensure that they understand the method and the way of training to support their trainers in the process.

> The Learning Center consultant convinced me to participate in a 10-hour class in Job Instruction together with the leaders on my leader team. Here I realized that it is a smart learning method and I must admit that when I saw the Fire Underwriters' Knot being demonstrated, it also made a big impression on me. In the Job Instruction class, I used an escalation process as my job for demonstration. This is a process that I was really struggling with getting people to understand and it turned out that I myself was also not clear on the steps for this job. By breaking it down together in the team, though, they became clear to me.
>
> One of my leaders brought in a job on how to follow up on costs in our financial system and during the class it became very clear how to do this. This was an eye-opener for me as I realized what could be obtained by this method from these two very clear and specific examples. The leadership team got a good understanding of how the method can contribute and we got a common picture of the methodology and what it takes to make the training material. It requires a real willingness to do it, but change is always difficult. The value in the method is that if you have specific

> problems to solve, for instance requiring a certain level of quality that must be trained using the method, it ensures that you almost always will remember it forever. So if you learn it once, you will remember it. That is a very important part.
>
> Another important element is the three areas defined in the method: the Important Steps, the Key Points and the Reasons for the Key Points. If you break down a job in the right way there will only be a small handful of Important Steps with some Key Points and their related reasons. So it is a very simple structure of the job. While the method should not be used for all jobs, the advantage is that this makes it possible to perform the job in a standardized way thereby making it easier to scale to other production sites. In addition to the scalability, it is also impacting the quality of the work we do. We get a more skilled workforce by learning and building on each other's competencies and this improves the quality of the work. The jobs where this is important in my area are critical parts of the process to ensure the right quality and this is where it creates great value.

When the trainers in the area started working with the TWI method, they chose to practice on a complex job with which they were struggling. This job was a critical part of the process and done with a high level of precision. Furthermore, it turned out that this part of the process was performed differently from one worker to another. The job was fixing the mold part in the milling machine. After the trainers had completed the 10-hour Job Instruction class, we started practicing the training methodology on this job and breaking it down for instruction.

> The first job we used to try out the method was the way we prepare and fix the mold part into the CNC milling machine. This is a very important part of the process. I have one employee and he has the ability to make his molds correctly the first time, every time as he has higher precision in his work and a higher degree of First-Time-Through (FTT). He is working on the same machines, the same programs, and with the same tools as his colleagues and, even though all those factors are the same, the quality of his work is higher than the others. The purpose was to find out what this guy was doing to achieve this level of performance. There is no doubt that when you do this you really are digging into the

inner core of the craftsmanship of the toolmaker and the details of how they work.

But this created some resistance in the organization, so it was important as a leader to create buy-in for the process. The way the TWI implementation has influenced us as an organization is by acknowledging that we have to describe how we work and that there is a big value in aligning our work. From a global perspective, it makes it much easier to transfer knowledge from one production site to another. There is no doubt about that.

In Mexico, the full process of wire cutting and spark erosion was an important task to break down and be able to train in a standardized way. The training methods used were a combination of classroom training and Job Breakdowns. One global trainer and one technical writer worked intensively providing the material for the jobs and trying out the Job Breakdowns on their colleagues in Billund, who also had a need for training.

We have specifically been working on describing our processes for wire cutting and spark erosion which are quite complex processes. This has been done in great detail and this makes it possible for Mexico to run these processes by themselves and they are able to train others based on this. We have seen that the description of the job is fundamental, otherwise you cannot do it. And they can now actually work on these technologies with a relatively small amount of knowledge prior to the training. This is a great success. It takes resources to set-up the descriptions, but once this is done it is much easier to train new people.

One example in my area where we have used TWI to great benefit is the job of defining how to do the programming for the CNC machines. In a very short time I can now explain the way to do this and I can do the process confirmation myself. As leader this makes it easy for me to set a clear direction on how the job should be done and to standardize the way it is done. Recently we had a situation with some quality issues on this job and I chose to bring it to the table in a department meeting emphasizing the process of the job. It was very easy for me to emphasize the direction for doing this job using the materials I had made for the training.

The process of breaking down the jobs was done by some of the specialists in the area with coaching from the Learning

Center consultant. We have good experience with mixing groups of experienced and less experienced employees to set this up and the results have been really fantastic. The people who were part of defining the breakdowns feel ownership of the process and had a positive experience, even though most of them indicated at the beginning that it was impossible to standardize this. In general, though, we meet some resistance the first time people experience this method, the same resistance that I had been through myself in the beginning. I think a way to loosen up this resistance could be to not only train the method on simple jobs in the beginning but also show a complex process that can be described in a very simple way. This could help in creating a more diverse picture of the value in the method.

We have now defined four to five of these strategies within our processes and we used them for training both existing employees to align the way we work and, naturally, new employees. It has been very exciting. This has a high value and probably the highest results we have obtained to date. By standardizing the work we have saved at least 1 million Danish Crowns through rework reduction and quality assurance by reducing the complexity in the tasks and ensuring a holistic way of describing them from A to Z. And by supporting the employees in solving their tasks more correctly and more efficiently, the complexity reduction is an advantage for the employee, the trainer and the leader.

The way this has influenced my organization is a change from craftsmanship towards a more industrial and professional approach in our working processes. I can find many examples where we do not have a strategic description supporting the processes and by providing this we can save a lot. There is a tendency to think that the TWI method is only for production and hourly paid work. But there are many other administrative areas, for instance IT [information technology] and planning activities or technical environments, where this method can be used. I have no doubt about that. It is a very simple way of explaining which process to go through, what to do, and in which sequence, how and why. I think that by providing this structure it is an easier way to remembering, so it is about learning an easy way.

Index

J

K

L

V

W